T0310329

FUNDAMENTALS OF COGNITIVE RADIO

Wiley Series on

Adaptive and Cognitive Dynamic Systems: Learning, Signal Processing, Communications, and Control

Series Editor: Simon Haykin
McMaster University, Ontario, Canada

This series of books, updated on its previous title *Adaptive and Learning Systems: Signal Processing, Communications, and Control*, has been renamed to make room for cognition, inspired by the human brain.

To be more specific, cognition is a powerful new way of thinking about a variety of applications, exemplified by Cognitive Radio, Cognitive Control, and Cognitive Radar, just to name a few. It has not only enriched pervasive disciplines such as leaning machines, signal processing, communications, and control, but it has also provided the means to bring to the table new applications and performance improvements not achievable by traditional means.

Another topic that is becoming increasingly important is networks. In this context, adaptive networks and cognitive networks are featuring in various applications that involve the environment, radar, radio, the power grid, etc.

Speaking of Cognitive Dynamic Systems, here again inspired by the brain, we may build on such systems and think of them as platforms for a new generation of Intelligent Decision-Making Computers that provide the machinery for supervising networks that are challenging, exemplified by the following three applications:

- Cognitive radio networks embedded into the world of wireless communication providers for improved radio spectrum utilization.
- Cognitive power-grid networks for improved utilization of traditional sources of energy and new sources involving photovoltaic systems for harnessing radiation from the sun, and powerful turbines for harnessing the power of winds.
- Cognitive radar networks for national security within as well as at borders.

To summarize, what we have described here is a new generation of ideas, and brand new applications that involve learning theory, signal processing, information processing, communication theory, control theory, and Monte Carlo simulations.

Related Titles

Bayesian Signal Processing: Classical, Modern, and Particle Filtering Methods, 2nd Edition
by James V. Candy

Data-Variant Kernel Analysis
by Yuichi Motai

Radio Resource Management in Multi-Tier Cellular Wireless Networks
by Ekram Hossain, Long Bao Le, DusitNiyato

Neural-Based Orthogonal Data Fitting: The EXIN Neural Networks
by GiansalvoCirrincione, Maurizio Cirrincione

Multiple-Input Multiple-Output Channel Models: Theory and Practice
by Nelson Costa, Simon Haykin

Adaptive Signal Processing: Next Generation Solutions
by TülayAdali, Simon Haykin

Kernel Adaptive Filtering: A Comprehensive Introduction
by Weifeng Liu, Jose C. Principe, Simon Haykin

Handbook on Array Processing and Sensor Networks
by Simon Haykin, K. J. Ray Liu

Complex Valued Nonlinear Adaptive Filters: Noncircularity, Widely Linear and Neural Models
by Danilo P. Mandic, Vanessa Su Lee Goh

Space-Time Layered Information Processing for Wireless Communications
by MathiniSellathurai, Simon Haykin

Knowledge Based Radar Detection, Tracking and Classification
by Fulvio Gini, MuralidharRangaswamy

Correlative Learning: A Basis for Brain and Adaptive Systems
by Zhe Chen, Simon Haykin, Jos J. Eggermont, Suzanna Becker

Adaptive Approximation Based Control: Unifying Neural, Fuzzy and Traditional Adaptive Approximation Approaches
by Jay A. Farrell, Marios M. Polycarpou

Model-Based Signal Processing
by James V. Candy

Acoustic Echo and Noise Control: A Practical Approach
by Eberhard Hänsler, Gerhard Schmidt

Least-Mean-Square Adaptive Filters
by Simon Haykin (Editor), B. Widrow (Editor)

Adaptive Control Design and Analysis
by Gang Tao

Stable Adaptive Control and Estimation for Nonlinear Systems: Neural and Fuzzy Approximator Techniques
by Jeffrey T. Spooner, Manfredi Maggiore, RaúlOrdóñez, Kevin M. Passino

Intelligent Image Processing
by Steve Mann

Kalman Filtering and Neural Networks
by Simon Haykin

FUNDAMENTALS OF COGNITIVE RADIO

Peyman Setoodeh
Simon Haykin

WILEY

This edition first published 2017

Registered Offices
John Wiley & Sons, Inc., 111 River Street, Hoboken, NJ 07030, USA

Editorial Office
111 River Street, Hoboken, NJ 07030, USA

For details of our global editorial offices, customer services, and more information about Wiley products visit us at www.wiley.com.

Wiley also publishes its books in a variety of electronic formats and by print-on-demand. Some content that appears in standard print versions of this book may not be available in other formats.

Library of Congress Cataloguing-in-Publication Data

Names: Setoodeh, Peyman, 1974- author. | Haykin, Simon S., 1931- author.
Title: Fundamentals of cognitive radio / Peyman Setoodeh, Simon Haykin.
Description: Hoboken, NJ : John Wiley & Sons, 2017. | Includes
bibliographical references and index.
Identifiers: LCCN 2017004056 (print) | LCCN 2017010420 (ebook) | ISBN
9781118302965 (cloth) | ISBN 9781119405832 (Adobe PDF) | ISBN
9781119405849 (ePub)
Subjects: LCSH: Cognitive radio networks. | Wireless communication systems.
Classification: LCC TK5103.4815 .S48 2017 (print) | LCC TK5103.4815 (ebook) |
DDC 621.384–dc23
LC record available at https://lccn.loc.gov/2017004056

Cover image: (Top image) © enjoynz/Gettyimages;
(Bottom image) © Andy Dean/Gettyimages
Cover design by Wiley

Set in 10/12pt TimesLTStd by SPi Global, Chennai, India

Printed in the United States of America

10 9 8 7 6 5 4 3 2 1

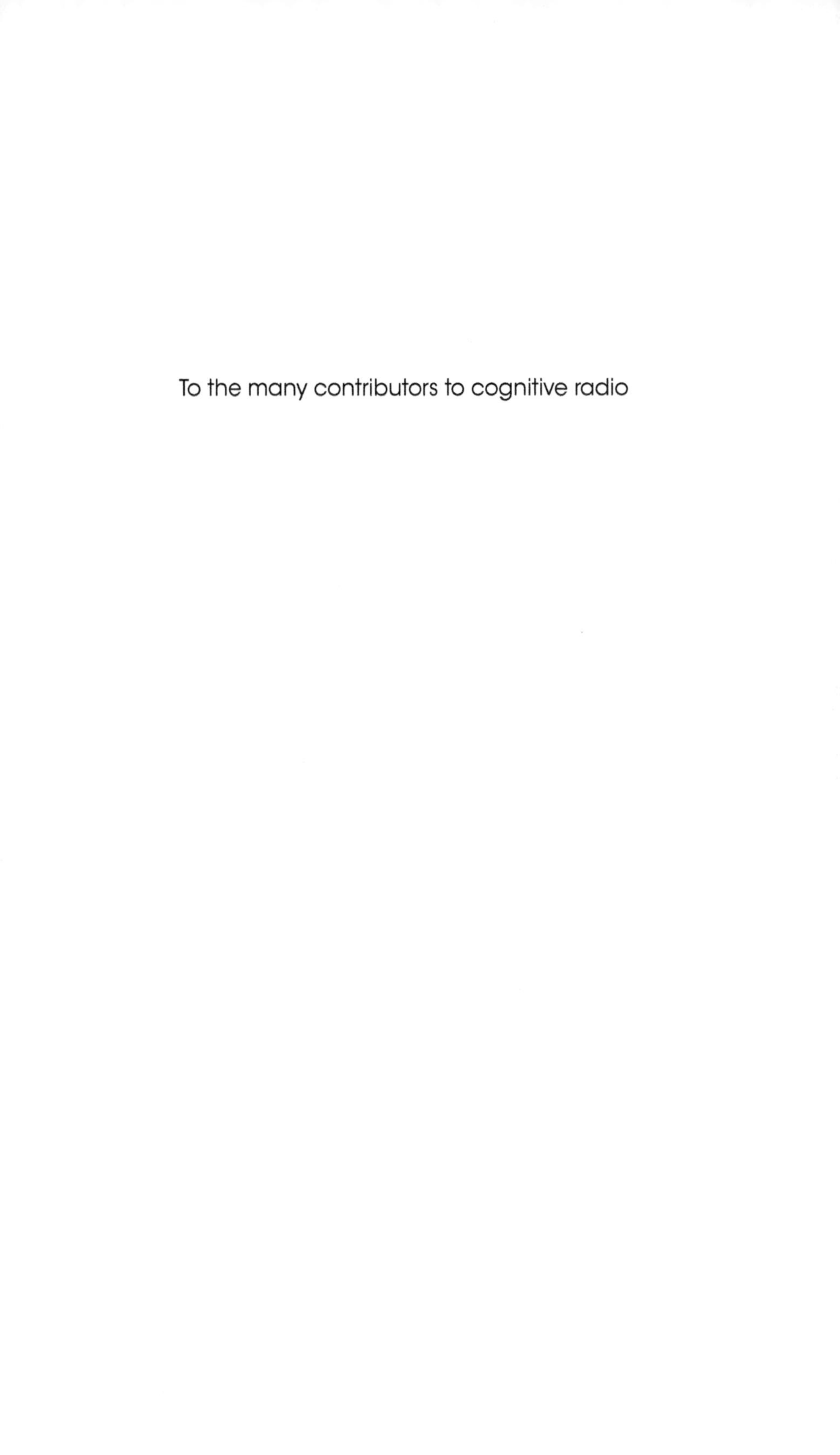

To the many contributors to cognitive radio

CONTENTS IN BRIEF

CONTENTS

LIST OF FIGURES

LIST OF TABLES

PREFACE

This book provides a new way of thinking on *cognitive radio networks*, proceeding beyond the traditional viewpoint. Cognitive radio provides a basis for addressing the practical issue of spectrum scarcity. This issue has been raised due to the continuing advances in wireless technology, which has led to ever-increasing demand for larger bandwidth. The issue of spectrum scarcity has been exacerbated due to inefficient utilization of the electromagnetic spectrum. The novel idea of cognitive radio is adopted for secondary usage of underutilized subbands. This leads to the existence of two worlds of wireless communications going on side by side: the legacy wireless world and the cognitive wireless world. Spectrum holes (i.e., the unused spectrum subbands) are the medium through which these two worlds dynamically interact. Releasing subbands by primary users allows the cognitive radio users to sustain communication and perform their normal tasks. Combination of the two wireless worlds can be viewed as a spectrum-supply chain network, in which the legacy owners and their customers (primary users) play the role of the suppliers and cognitive radios (secondary users) play the role of consumers. This book discusses two classes of spectrum-supply chain networks based on two regimes: one allows *open access* to the spectrum and the other is a *market-driven* regime. Each one of them has its own merits and suitability for a different environment; therefore, they have complementary roles. After covering the basic building blocks of a cognitive radio transceiver, analytic models are developed for these two classes of networks, which pave the way for analysis of both equilibrium and transient behaviors.

In order to improve the efficiency and sustainability of the spectrum-supply chain network, an artificial economy is designed based on viewing the licensed and unlicensed bands as private goods and common-pool resources, respectively. The proposed framework addresses the issue of spectrum sharing across licensed and unlicensed bands. It aims for Pareto optimality by trying to achieve the Lindahl equilibrium. The proposed framework facilitates the integration of the two wireless worlds and paves the way for commercialization of cognitive radio.

Building on the developed spectrum-supply chain paradigm, an economic model is presented for heterogeneous networks (HetNets), which captures their multitier nature. HetNets provide a way for enhancing the spectral efficiency via spatiotemporal reuse of spectrum. The HetNet paradigm appears to be a main pillar for 5G.

The proposed economic model is based on decoupling of spectrum and network infrastructure. Here, the problem of resource sharing among networks whether it be spectrum, infrastructure, or both is formulated as the network horizontal merger at different levels. It allows for addressing diverse issues such as green communication and spectral efficiency as well as ubiquitous networking and services.

Peyman Setoodeh and Simon Haykin

Hamilton, Ontario, Canada
September 2016

ACKNOWLEDGMENTS

We would like to express our deepest gratitude to Dr Timothy N. Davidson, Dr David Earn, and Dr Matheus Grasselli, McMaster University, Hamilton, Ontario, Canada, for the fruitful discussions that led to many ideas in this book. Special thanks goes to Dr Farhad Khozeimeh, Amazon, Inc., for helping with the section on dynamic spectrum management and Dr David J. Thomson, Queen's University, Kingston, Ontario, Canada, and Dr Jeffrey H. Reed, Virginia Tech., Blacksburg, Virginia, USA, for helping with the section on spectrum sensing. We also wish to thank Dr Tamás Terlaky, Lehigh University, Bethlehem, Pennsylvania, USA, for many useful suggestions. Sincere thanks to Dr Sergio Barbarossa, Sapienza University of Rome, Rome, Italy, for his valuable comments on different aspects of cognitive radio networks. We are grateful to Dr Keith E. Nolan, Intel Corporation, Dr James O. Neel, and Dr Ryan Leduc, McMaster University, for helping in the early stages of this research.

We also owe many thanks to our friends and former colleagues in the Cognitive Systems Laboratory, McMaster University; Dr Yanbo Xue, D-Wave Systems, Inc.; Dr Ulaş Güntürkün, Ultra Maritime Digital Communications Center; Dr Jiaping Zhu, BMO; Dr Karl Wiklund, Vitasound Audio; Mr Kenny Szeto, Technical Solutions Engineering at Turn; Ms Mathangi Ganapathy, TÜV SÜD; Dr Nelson Costa, Cognitive Systems Corp.; Dr Mehdi Fatemi, Maluuba, Inc.; Dr Ashkan Amiri, RBC; Dr Ienkaran Arasaratnam, Ford Motor Company; Dr Patrick Fayard, Dr Adhithya Ravichandran, Mr Jerome Vincent, Mr Kevin Kan, Mr Shuo Feng, and Mr David Findlay for their kindness, friendship, valuable help, and suggestions.

We sincerely thank Mrs Lola Brooks, Ms Rachel Harvey, Ms Laura Kobayashi, Dr Bartosz Protas, Mr Terry Greenlay, Mr Cosmin Coroiu, Mrs Cheryl Gies, and Mrs Helen Jachna, who were always welcoming, supportive, and helpful.

The detailed feedback notes on different aspects of the book by reviewers have not only reshaped the book into its present form but also made the book the best it could be.

Finally, we would like to thank our families. Their endless support, encouragement, and love have always been a source of energy for us.

Peyman Setoodeh and Simon Haykin

ACRONYMS

4G	fourth generation wireless systems
5G	fifth generation wireless systems
AR	autoregressive
ATSC	Advanced Television Systems Committee
AVI	affine variational inequality
AWGN	additive white Gaussian noise
BPSK	binary phase-shift keying
CAPL	channel allocation priority list
CDS	cognitive dynamic systems
CFDP	cycle frequency-domain profile
CPS	cyber-physical systems
CR	cognitive radio
DLD	double-layer dynamics
DOA	direction of arrival
DSL	digital subscriber line
DSL	digital subscriber lines
DSM	dynamic spectrum management
DTV	digital television
EMA	exponential moving average
EMA	exponential moving average
ESS	evolutionary stable strategy
EVI	evolutionary variational inequality
FBMC	filter bank multicarrier
FCC	Federal Communications Commission
FDE	functional differential equation
FFT	fast Fourier transform
FFTW	the fastest Fourier transform in the west

GMSK	Gaussian minimum-shift keying
GSM	Global System for Mobile communication
HDS	hybrid dynamic system
HetNets	heterogeneous networks
HMM	hidden Markov model
ICI	inter-carrier interference
IEEE	Institute of Electrical and Electronics Engineers
IFFT	inverse fast Fourier transform
IoS	Internet of services
IoT	Internet of things
ISI	inter-symbol interference
IWFA	iterative waterfilling algorithm
KKT	Karush–Kuhn–Tucker
LCP	linear complementarity problem
LTE	long-term evolution
M2M	machine-to-machine
MG-DSM	minority game-based dynamic spectrum management
MIMO	multiple-input multiple-output
MLCP	mixed linear complementarity problem
MSK	minimum-shift keying
MTM	multitaper method
NCP	nonlinear complementarity problem
NFV	network function virtualization
NTU	nontransferable utility
ODE	ordinary differential equation
OFDM	orthogonal frequency division multiplexing
OFDMA	orthogonal frequency division multiple access
OQPSK	offset quadrature phase-shift keying
PAM	pulse-amplitude modulation
PAPR	peak-to-average power ratio
PDA	personal digital assistant
PDS	projected dynamic system
PDSD	projected dynamic system with delay
PSWF	prolate spheroidal wave function
PU	primary user
PWA	piecewise affine
QoS	quality of service
QPSK	quadrature phase shift keying
REM	radio environment map

RF	radio frequency
RFID	radio-frequency identification
RKRL	radio knowledge representation language
RSA	radio scene analysis
RX	receiver
SCM	single-carrier modulation
SCM	single-carrier modulation
SDN	software-defined networking
SDR	software-defined radio
SINR	signal-to-interference plus noise ratio
SNR	signal-to-noise ratio
SO-DSM	self-organized dynamic spectrum management
SOM	self-organizing map
SU	secondary user
SVD	singular value decomposition
TF-MSC	two-frequency magnitude-squared coherence
TFA	time-frequency analysis
TPC	transmit-power control
TU	transferable utility
TX	transmitter
VI	variational inequality
VPC	virtual power cube

1

INTRODUCTION

1.1 THE FOURTH INDUSTRIAL REVOLUTION

The *fourth industrial revolution*, which is denoted by the term "Industry 4.0," is in its early stages. The hallmarks of the former three industrial revolutions are as follows:

- Deployment of mechanical production facilities
- Use of electric power for mass production and communications
- The digital revolution

The distinct feature of Industry 4.0, which distinguishes it from its three predecessors, is the fact that it has been predicted a priori instead of being observed by postanalysis. This prediction opens a window of opportunity for futurists as well as visionary individuals and institutes to actively participate and play key roles in engineering the future. Industry 4.0 is built upon the following four pillars [1]:

- *Cyber-physical systems* (*CPS*), which include *smart products* as their subcomponents
- *Internet of things* (*IoT*), which relies on *machine-to-machine* (*M2M*) *communication* as an enabling technology
- *Internet of services* (*IoS*), which is exemplified by *cloud computing* as a model for allowing Internet-enabled devices to access a shared pool of configurable computing resources according to their needs
- *Smart factories*.

Fundamentals of Cognitive Radio, First Edition. Peyman Setoodeh and Simon Haykin.

Industry 4.0 will produce *massive* amounts of data. Portions of the produced data that are associated with high volume, variety, and velocity (3 Vs) are referred to as *big data*. Each one of the above pillars will be briefly described in what follows [2–7].

By definition, the integration of digital computing and a physical environment results in a cyber-physical system. Therefore, many applications can be collected under the umbrella of such systems [8]. A cyber-physical system usually includes a distributed set of different sensors, where the number of sensors depends on the scale of the system. Data gathered by these sensors are used to form a representation of the environment, which is then used for decision-making. Only portions of the gathered data will be useful (i.e., relevant to) the decision-making task. Hence, in accordance with the task at hand, the information extracted from the available data can be divided into two sets: relevant and irrelevant. The former provides the *actionable information* [9].

In order to perform a task with an acceptable level of risk, a specific amount of information is required, which is called sufficient information. If the actionable information does not meet the information sufficiency criteria from the decision-making perspective, the decision-maker will face an *information gap*. In Ref. [10], *cognitive control* was proposed to reduce this gap between the actionable and sufficient information sets by controlling the directed flow of information:

> Given a probabilistic dynamic system that includes a perception–action cycle, and ideally mimics the human brain, the function of cognitive control is to adapt the directed flow of information from the perceptual part of the system to its executive part so as to reduce the information gap, which is equivalent to reducing a properly defined risk functional for the task at hand, the reduction being with a probability close to one.

In a cyber-physical system, cognitive and physical controllers play complementary roles. Cognitive complementary actions can influence different parts of the system:

- Cognitive actions may be applied to the environment in order to indirectly affect the perception process.
- Cognitive actions may also be applied to the system itself in order to reconfigure the sensors and/or actuators.
- In addition, cognitive actions may also be subsumed in physical-control actions. In such a case, a physical action is applied to the system but with the goal of decreasing the information gap (with or without other goals).

In other words, in large-scale CPS with the requirement of critical decision-making, cognitive control will enhance reliability of the decision-making process. Final decisions (i.e., outputs of the decision-making process) are then sent as a set of commands to different actuators, which are also distributed in the system. As a result, CPS have been significantly transforming the way our modern society perceives the physical world and interacts with it.

The word "thing" in the Internet of Things refers to different entities, such as radio-frequency identification (RFID), sensors, actuators, computers, and mobile wireless devices, which may all be smart products that communicate with each other using a specific addressing scheme and cooperate toward a common goal. "Thing" can also be interpreted as cyber-physical entities and in effect, therefore, IoT can be viewed as a network of CPS [1].

Adopting a systematic viewpoint on organizations has led to the idea of *value chain*, which refers to activities that are performed by an organization in order to deliver a valuable product or service to market [11]. Regarding the fact that CPS have made the fusion of physical and virtual worlds possible [1], a combination of both physical and virtual value chains must be taken into account. Hence, a service-oriented architecture can be created that promotes distributed production control, which is built on modular assembly stations that are connected by automated guided vehicles. This provides customers with special production technologies and gives them a degree of freedom to somehow custom design the product that they need. Customers can use the assembly stations and the associated automated transportation system through the IoS.

A factory is said to be smart if in a context-aware manner it can help both employees and machines to execute the tasks that are assigned to them. The operation of such factories would be demand-driven. As proposed in [12], what distinguishes a smart factory from an ordinary one is the existence of the so-called *calm systems* that operate in the background and are able to communicate and interact with their environments. A smart factory can be viewed as a factory, in which CPS communicate over the IoT to facilitate execution of assigned tasks to employees and machines [1]. The idea of calm systems is very similar to the notion of *cognitive dynamic systems* (CDS) [13], which can play the role of the central nervous system for the smart factory.

It is obvious that all the mentioned pillars of Industry 4.0 rely on communications and networking in one way or another. In this regard, *cognitive radio* as a smart product will be part and parcel of each one of the pillars. Therefore, it can be expected that cognitive radio networks will play essential roles in the years to come far beyond the initially set mission for improving spectral efficiency. In order to exploit the full potential of cognitive radio networks in light of the fourth industrial revolution, such networks must be studied as *intelligent sociotechnical systems* [14]. Therefore, the socioeconomic principles for self-organizing institutions [15] provide guidelines for design and analysis of cognitive radio networks, where decentralized control, competition for limited resources, and vulnerability to both intentional and unintentional errors, are the main characteristics. In such an environment, the emphasis should be on the endurance of the resource-distribution mechanism rather than on its optimality. According to [16], eight design principles must be considered for endurance of self-management of common-pool resources:

1. Clearly defined boundaries for determining who has the right to use which portion of the resources.
2. Congruence between appropriation and provision rules and the state of the prevailing local environment.
3. Collective-choice arrangements.
4. Monitoring of both state conditions and users' behavior.

5. A flexible scale of punishment for users that violate communal rules.
6. Access to fast, cheap conflict-resolution mechanisms.
7. Existence of and control over their own institutions without intervention by external authorities.
8. Systems of systems as layered or encapsulated common-pool resources, with local resources at the base level.

Regarding the potentials of cognitive radio networks to play key roles in Industry 4.0 especially, in implementing intelligent sociotechnical systems, the primary objective of this book is to introduce a relatively new way of thinking about such networks. With this objective in mind, we find it instructive to start the discussion with the spectrum-supply chain paradigm, which may not be that well known in the signal-processing and communication-systems communities.

The notion of a *supply chain* may be viewed as follows: A supply chain is made up of different entities, the role of which is to fulfill the request of a specific customer. In addition to manufacturers, suppliers, and customers, a supply chain includes transporters, warehouses, brokers, and retailers. A supply chain also includes all functions of each party such as new product development, marketing, operations, distribution, finance, and customer service. Supply chains have been gradually evolving to complex structures known as *supply chain networks*, which are characterized by their dynamic nature due to continuous flow of information, product, and funds between different parties [17].

Supply chain networks including energy networks, food and water-resource networks, transportation networks, and communication networks play critical roles in shaping and survival of our modern-day society. Hence, it is significantly important to understand how an efficient supply chain must be designed and managed, how weaknesses can be improved, and how each party should plan and operate in order to achieve a competitive advantage [18].

Both urban and rural environments have been transforming by new emerging technologies. Even social relevance has been affected by these technologies. One of these transforming technologies is the wireless technology, which has manifested itself as a revolutionary paradigm shift through applications that have already gone far beyond our imagination. The applications include, but not limited to, multimedia communications, telemedicine, sensor networks, smart spaces (e.g., homes, offices), and recently smart cities [19]. This book revisits wireless communications in the light of recent understandings of supply chain networks.

1.2 COGNITIVE RADIO

The second great unification in physics is attributed to Maxwell for integrating electricity, magnetism, and optics under one unifying umbrella, collectively named the theory of classic electromagnetism [20].[1] However, the original formulation, which was published in 1865, included a set of 20 equations instead of the celebrated 4 partial differential equations known as Maxwell's equations [21]. Later on, in 1873, Maxwell presented his theory in its final form in a two-volume book [22].[2]

In 1887, Hertz experimentally demonstrated the physical existence of radio waves, which had been predicted by Maxwell about 20 years earlier [24, 25]. This historical achievement confirms the following statement: Theory and experiment must go hand in hand to flourish in their own respective ways. After his successful experiment, Hertz said:

> I do not think that the wireless waves I have discovered will have any practical application.

Despite this prediction, mobile communications and the broadband Internet access have been playing two key roles in the development of our society in recent years: (1) The increasing number of users of Internet-enabled wireless devices illustrates the shift away from traditional application-specific radio technology to service-oriented information delivery systems. (2) Regarding the ever-increasing demand for more advanced applications that require transferring higher volumes of data, communication technologies are progressing toward providing secure and seamless connectivity of mobile devices to any network, anytime, and anywhere [26, 27].

In this regard, to allocate the spectrum dynamically and openly, future wireless devices need to be service-oriented terminals, which are compatible with computer systems and support the unlocked and multiple wireless standards [28]:

> Spectrum is like air; we need to keep it clean, open, and green for our environment.

A significant emerging approach to improve spectrum utilization is through *temporal spectrum reuse*. *Cognitive radio* offers a novel way for improving the efficiency of spectrum utilization [29–31]. Deployment of scaled-down wireless base stations (i.e., *femtocells*), which use licensed spectrum belonging to an operator improves capacity and coverage through *spatial reuse of spectrum* [32]. However, conventional small cells put no new spectrum into play. Hence, a new paradigm for wireless communications that can harness the potential possibilities offered by these two different approaches and unify them under one umbrella would be appealing. Small-cell networks equipped with cognition (*cognitive small-cell networks*) will pave the way for efficient spectrum sharing across networks via *spatiotemporal reuse of spectrum* [33].

Interest in a new generation of engineering systems enabled with cognition, started with *cognitive radar* [34] followed by *cognitive radio*, a term that was coined by

[1]Newton's theory of classic mechanics is considered to be the first great unification in physics, which unified terrestrial and celestial mechanics [20].

[2]Heaviside takes credit for reducing the number of Maxwell's equations to make the theory more understandable for his contemporary scientists so that its importance could be appreciated [23].

Mitola and Maguire [35]. A cognitive system is built on five pillars: perception–action cycle, memory, attention, intelligence, and language [13, 36]. In [35], the idea of cognitive radio was introduced within the software-defined radio (SDR) community. Subsequently, Mitola elaborated on a so-called "radio knowledge representation language" in his own doctoral dissertation [29].

Furthermore, in a short section entitled "Research Issues" at the end of his doctoral dissertation, Mitola went on to say the following [29]:

> How do cognitive radios learn best? merits attention. The exploration of learning in cognitive radio includes the internal tuning of parameters and the external structuring of the environment to enhance machine learning. Since many aspects of wireless networks are artificial, they may be adjusted to enhance machine learning. This thesis did not attempt to answer these questions, but it frames them for future research.

Haykin in his seminal journal paper on cognitive radio, proposed the following definition [30]:

> Cognitive radio is an intelligent wireless communication system that is aware of its surrounding environment (i.e., outside world), and uses the methodology of understanding-by-building to learn from the environment and adapt its internal states to statistical variations in the incoming radio frequency (RF) stimuli by making corresponding changes in certain operating parameters (e.g., transmit-power, carrier-frequency, and modulation strategy) in real-time, with two primary objectives in mind:
>
> - highly reliable communications whenever and wherever needed;
> - efficient utilization of the radio spectrum.

Then, he provided a detailed exposition of signal processing, control, learning and adaptive processes, and game-theoretic ideas that lie at the heart of cognitive radio [30]. As shown in Figure 1.1, three fundamental cognitive tasks, embodying the perception–action cycle of cognitive radio, were identified in that 2005 paper [30]:

- Radio-scene analysis (RSA) of the radio environment performed in the receiver
- Transmit-power control and dynamic spectrum management (DSM), both performed in the transmitter
- Global feedback, enabling the transmitter to act and, therefore, control data transmission across the forward wireless (data) channel in light of information about the radio environment fed back to the transmitter by the receiver. In other words, information on spectrum holes (i.e., underutilized subbands) and the forward channel's condition, extracted by the scene analyzer at the receiver, is sent to the transmitter via a feedback channel.

In effect, the emphasis in that 2005 paper was placed on cognitive radio as a "closed-loop feedback control system" with practical benefits and the need for precautionary measures, recognizing that feedback is a "double-edged sword".

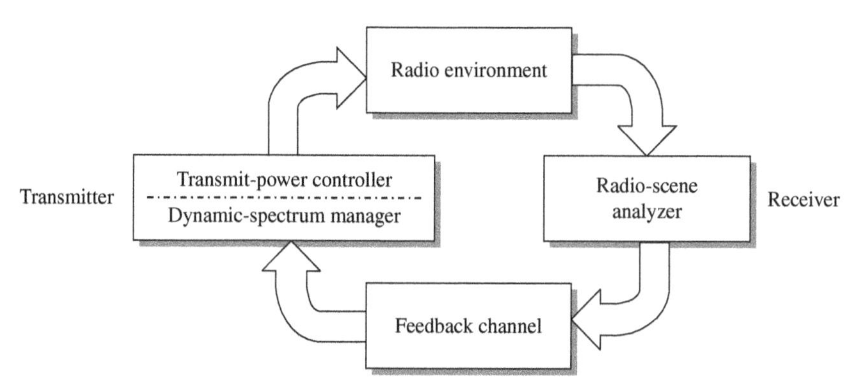

FIGURE 1.1 The cognitive information-processing cycle in cognitive radio. A cognitive radio transceiver is built on a perception–action cycle. Radio-scene analyzer in the receiver plays the role of the perceptor. Dynamic spectrum manager and transmit power controller in the transmitter play the role of the executive part. The perceptual and executive parts together with the feedforward and feedback channels form a closed-loop system. Source: Haykin and Setoodeh (2015) [37]. Reproduced with the permission of IEEE.

Since its inception over a decade ago, interest in cognitive radio, its theory and applications has grown exponentially. The driving force behind this exponential growth is summed up as follows [13]:

> Cognitive radio has the potential to mitigate the radio-spectrum underutilization problem in today's wireless communications.

With this engineering challenge in mind, we begin the study of cognitive radio with this critical issue as our starting point.

1.3 THE SPECTRUM-UNDERUTILIZATION PROBLEM

The electromagnetic *radio spectrum* is a natural resource, the use of which by transmitters and receivers is licensed by governments. In November 2002, the Federal Communications Commission (FCC) published a report prepared by the Spectrum-Policy Task Force aimed at improving the way in which this precious resource is managed in the United States [38]. The task force was made up of a team of high-level, multidisciplinary professional FCC staff: economists, engineers, and attorneys – from across the commission's bureaus and offices. Among the task force's major findings and recommendations, the second finding on page 3 of the report is rather revealing in the context of spectrum utilization [38]:

> In many bands, spectrum access is a more significant problem than physical scarcity of spectrum, in large part due to legacy command-and-control regulation that limits the ability of potential spectrum users to obtain such access.

Indeed, if we were to scan portions of the radio spectrum including the revenue-rich urban areas, we would find that

- some frequency bands in the spectrum are largely unoccupied most of the time;
- some other frequency bands are only partially occupied; and
- the remaining frequency bands are heavily used.

The underutilization of the electromagnetic spectrum leads us to think of a new term commonly called *spectrum holes*, for which we offer the following definition [13, 30]:

> A spectrum hole is a band of frequencies assigned to a primary (licensed) user, but, at a particular point in time and specific geographic location, the band is not being utilized by that user.

Spectrum utilization can be improved significantly by making it possible for a secondary (cognitive radio) user (who is not being serviced) to access a spectrum hole unoccupied by the primary (legacy) user at the right location and time in question.

Cognitive radio offers a new way of thinking on how to promote efficient use of the radio spectrum by exploiting the existence of spectrum holes. In a related context, the *spectrum-utilization efficiency* of cognitive radio is assessed in the context of four practical system issues [13]:

1. *Accuracy and reliability*, with which the spectrum holes are identified.
2. *Computational speed*, with which the spectrum-hole identification is accomplished.
3. *Management of resources*, which involves the allocation of spectrum holes among competing secondary users in the cognitive radio network, effectively and reliably.
4. *Coexistence of the cognitive radio network alongside the legacy radio network*, which will have to be accomplished in a harmonious manner for the good of all users, both secondary and primary.

Requirements (1) and (2) are responsibilities of receivers in the cognitive radio network, while the transmitters in the network are responsible for requirement (3). As for requirement (4), with the legacy radio network having paid for using the spectrum and legally approved by regulatory agencies, the responsibility for this last requirement rests with the cognitive radio network, viewed as a system of systems [13].

1.4 COUNTRYWIDE MEASUREMENTS OF SPECTRUM UTILIZATION

According to predictions made by the International Telecommunications Union as well as the Organization for Economic Cooperation and Development, unless serious actions are taken toward smart, efficient, and dynamic management of the electromagnetic spectrum, the worldwide mobile communication network would face serious problems in the near future [28].

In this context, several measurement studies have been conducted in different countries around the world, as summarized in Table 1.1.

In the United States, measurements have shown that from January 2004 to August 2005, on average, only 5.2% of the radio spectrum was actually in use [39]. Measurements over a period of 2 days in November 2005 showed that the average spectrum occupancy in the band 30–3000 MHz was 13.1% and 17.4% for New York and Chicago, respectively [40]. In [42], the spectrum occupancy in the band 400–7200 MHz was compared for an urban area (Atlanta, Georgia) and a rural area (North Carolina); the respective measurements were 6.5% and 0.8%. At the Loring Commerce Centre, Limestone, Maine, USA, measurements over a period of 3 days in the band 30–3000 MHz, showed that the average spectrum usage was 1.7%; occupancy varied from less than 1–24.65% in different subbands. The maximum occupancy of 24.65% was reported for the band 470–512 MHz [41].

In Auckland, New Zealand, the spectrum occupancy was reported to be 6.2% over the frequency band 806–2750 MHz [43].

In Singapore, the average spectrum occupancy in the band 80–5850 MHz, based on measurements over a period of 12 days, was reported to be 4.54% [44].

In Doha, Qatar, measurements performed over a period of 3 days in the 700–3000 MHz frequency band showed that the spectrum utilization was 1% for indoor environments and 15.3% for outdoor environments [45].

It is worth mentioning that the results just mentioned highly depend on sensing locations, the spectrum sensing method, and the chosen threshold used to distinguish idle bands from occupied bands.

1.5 WHY BE INTERESTED IN COGNITIVE RADIO NETWORKS?

A significant emerging approach to improve spectrum utilization is through *temporal spectrum reuse*. *Cognitive radio* offers a novel way for improving the efficiency of spectrum utilization. Mitola's main idea described in [35] and [29] was to equip wireless personal digital assistants (PDAs) and related networks with a level of computational intelligence that they can detect users' communication needs and provide them with appropriate radio resources and wireless services. Haykin identified the basic building blocks of cognitive radio and developed a framework based on signal processing, communications, information theory, control theory, and game theory, which provides guidelines for implementing cognitive radio [30]. In this framework, the use of cognitive radio addresses the spectrum-utilization issue by identifying the underutilized subbands of the electromagnetic spectrum and then, providing the means for making those subbands available for employment by secondary users, who do not hold a license for using those subbands. Typically, the subbands allocated for wireless communications are the property of legally license owners, which, in turn, make them available to their own customers: the primary users. Naturally, the entire

TABLE 1.1 Sample Measurement Studies Regarding Spectrum Utilization in Different Countries

Country	Region in the Country	Frequency Range (MHz)	Usage (%)	Time	References
USA	–	–	5.2	January 2004–August 2005	[39]
	New York	30–3000	13.1	2 days in November 2005	[40]
	Chicago		17.4		
	Limestone	30–3000	1.7	3 days	[41]
	Atlanta	400–7200	6.5	–	[42]
	North Carolina (a rural area)	400–7200	0.8		
New Zealand	Auckland	806–2750	6.2	–	[43]
Singapore	Singapore	80–5850	4.54	12 weekdays	[44]
Qatar	Doha	700–3000	15.3	–	[45]

operation of cognitive radio hinges on the availability of spectrum holes that can be used for communication in an opportunistic manner. Moreover, the operation of cognitive radio is compounded further by the fact that the spectrum holes come and go in a rather stochastic manner, making the design of cognitive radio networks much more challenging.

1.6 DIRECTED INFORMATION FLOW

As described in [13], for a dynamic system to be cognitive, it has to embody four distinct processes: perception–action cycle, memory, attention, and intelligence. Every cognitive dynamic system has its own characteristic perception–action cycle, so does cognitive radio. However, before proceeding to address this issue, it is instructive to come up with a definition for what we mean by a "user" in a radio network. To this end, we first recognize that at each end of a wireless communication channel we have a *transceiver*, which embodies a transmitter and a receiver combined together as one whole unit. So, when we speak of a radio user, we offer the following definition [13]:

> A user refers to a communication link that connects the transmitter of a transceiver at one end of the link that is in communication with the receiver of another transceiver at the other end of the link. Moreover, the term "secondary user" is adopted for a cognitive radio, so as to distinguish it from the term "primary user", which is reserved for a legacy (i.e., licensed) radio unit.

Note the terms "primary user" and "secondary user" were used in the previous sections, ahead of this definition.

Referring to Figure 1.2, we see that on the right-hand side of the figure we have a receiver unit in the transceiver of cognitive radio (RX-CR) whose cognitive function is RSA, where CR stands for cognitive radio. On the left-hand side of the figure, at some remote location, we have a transmitter unit in the transceiver of cognitive radio (TX-CR) whose cognitive function is DSM and transmit power control (TPC) [47]. The RSA of the RX-CR unit senses the radio environment with the objective of identifying spectrum holes. This information is passed onto the TX-CR unit via the feedback channel. At the same time, through its own RSA, the TX-CR unit will have identified the spectrum holes in its own specific neighborhood. The combined function of the DSM and TPC in the TX-CR unit is to identify a spectrum hole that is common to it as well as the RX-CR unit, through which transmission over a data channel in the radio environment can be carried out. In this way, directed information flow across the cognitive radio is established on a cycle-by-cycle basis.

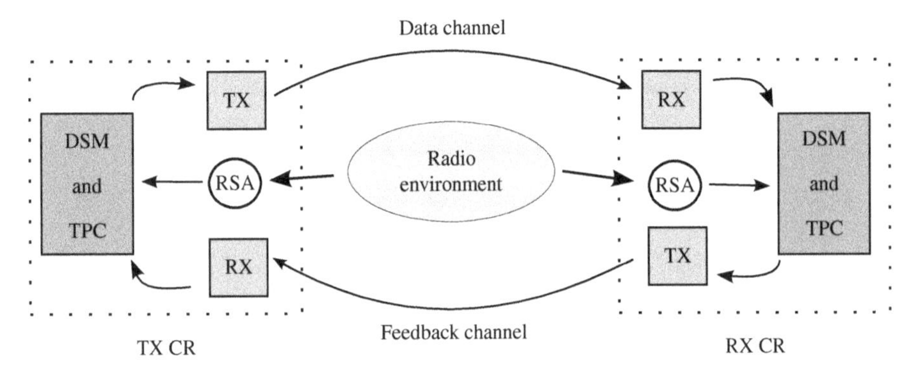

FIGURE 1.2 Directed-information flow in cognitive radio. DSM: dynamic spectrum manager; TPC: transmit-power controller; RSA: radio-scene analyzer; RX: receiver; TX: transmitter; TX CR: transmitter unit in the transceiver of cognitive radio; RX CR: receiver unit in the transceiver of cognitive radio. Source: Khozeimeh and Haykin (2009) [46]. Reproduced with the permission of John Wiley and Sons.

What we have just described is the very essence of the perception–action cycle for a communication link in cognitive radio. To add more specificity to the notion of RSA, "nonparametric spectrum estimation" is used in the receiver.

On the basis of Figure 1.2, we now address how the basic four processes of cognition are indeed satisfied, one by one. In so doing, we will have not only justified the rationale for radio cognition but also paved the way for the material to be covered in subsequent parts of the book [13].

1. *Perception–action cycle*: By using nonparametric spectrum estimation for perception in the receiver, the task of finding spectrum holes is achieved without having to formulate a "model" of the radio environment; hence, in effect, bypassing the need for perceptual memory. Spectrum estimation is an ill-posed inverse problem, which, therefore, requires the use of regularization. As will be explained later, the *multitaper method* (MTM) satisfies this requirement. Moreover, through the use of time-space processing, MTM provides the means for identifying the spectrum holes at a particular point in time as well as location in space. It is for these two reasons and a few others to be elaborated later on that we view the MTM as a method of choice for perception (i.e., spectrum sensing) of the radio environment.
2. *Learning and memory*: The task of DSM, to be discussed later, relies on the use of a learning process called *Hebbian learning*, inspired by the human brain [48]. An important characteristic of Hebbian learning is the inherent capability of self-organization. Thus, the dynamic spectrum manager has the practical means to dynamically choose and assign a set of appropriate links for communication to each cognitive radio user by learning the underlying environmental communication patterns. Knowledge thus learned about the communication patterns of the primary users in a radio network and, to some extent, those of

other secondary users in the local neighborhood is stored in memory. Moreover, the "synaptic" weights of a self-organized feature map are adaptively updated in response to new inputs, on a cycle-by-cycle basis.

Looking at Figure 1.2, we see the *coupling* between dynamic spectrum manager and the transmit-power controller. Specifically, through the use of *game-theoretic ideas*, also to be discussed later, and by virtue of information received from the nonparametric spectrum estimator through the feedback channel about interference levels in chosen feedforward communication channels, the transmit-power controller is enabled to *adaptively* adjust the transmitted radio signal, subject to prescribed constraints. In this *resource-allocation game*, the cognitive radio acquires the ability to reach equilibrium fast enough.

3. *Attention*: To illustrate how the process of attention manifests itself in cognitive radio, consider the following example. A serious accident has occurred at some particular point in time and specific location in space, thereby resulting in a "surge" in wireless-communication traffic. By virtue of built-in space-time processing, the nonparametric spectrum estimator sends information to the transmit-power controller, identifying which particular subbands of the radio spectrum have become congested due to the accident. Furthermore, in response to input from the radio-scene analyzer, the dynamic spectrum manager itself focuses its attention on those remaining subbands with lower interference levels. In so doing, communication over the newly found cognitive radio link is maintained, bypassing the congested subbands.

   Moreover, through its own self-organized learning process, the dynamic spectrum manager builds a *predictive model* of the radio environment. Using this model, the cognitive radio is enabled to predict the availability duration of spectrum holes, which, in turn, determines the predicted horizon of the transmit-power control mechanism.

4. *Intelligence*: As it is with human cognition, intelligence in cognitive radio builds itself on the processes of perception, memory, and attention, just described under points (1), (2), and (3), respectively. To appreciate the importance of intelligence, consider a cognitive radio network with multiple secondary users whose communication needs would have to be accommodated in a satisfactory manner. Accordingly, the perception – action cycle of Figure 1.2 would have to be expanded in a corresponding way, such that the secondary users share:

   - the radio environment for their individual forward communication needs and
   - separate wireless channel for their individual feedback requirements.

   In such a scenario, intelligence manifests itself as follows [13]:

   > Through a decision-making mechanism involving intelligent choices, the available resources (i.e., bandwidth and power) are equitably assigned to the secondary users in accordance with a prescribed protocol in the face of environmental uncertainties, and in such a way that the interference in the radio environment does not exceed a prescribed limit.

The environmental uncertainties include the reality that spectrum holes come and go in some stochastic manner, which may, therefore, mandate *robustification* of the transmit-power controller, an issue that is discussed later.

From the rationale just presented under points (1)-(4), it is now apparent that the four basic processes of cognition involved in radio for communication are satisfied. Now that it has become clear how cognitive capabilities are built into the system, the next step is to study issues pertaining to networks of cognitive radios.

1.7 COGNITIVE RADIO NETWORKS

A cognitive radio network, which is a *system of systems*, is a *goal-seeking* network in the sense described in [49]. The following classes of problems are involved in developing a cognitive radio network [37]:

- Specifying the goal that the system is pursuing (i.e., efficient spectrum utilization and ubiquitous network connectivity).
- Discriminating between the available alternative scenarios is based on the meaning of a desirable decision.
- Choosing a desirable action is based on a decision-making process.

By the same token, every system in the network (i.e., every cognitive radio) is a goal-seeking system too. *Game theory* has been extensively used to model complex interactions among different types of users with different desired payoffs [50]. In this context, researchers have investigated a wide range of problems and issues including but not limited to the following: spectrum sharing [51–53], scheduling [54], and interference managing [55].

For spectrum sharing, three different paradigms have been suggested in the literature [31]:[3]

- In the *underlay* paradigm, the interference caused by secondary-user transmitters on primary-user receivers must be kept below a certain threshold.
- In the *overlay* paradigm, the secondary users must have perfect knowledge about both primary and secondary users as well as the communication channels between them [56]. Sophisticated radio architectures are needed to use this knowledge and try to keep the interference caused by the secondary network on the primary network at the minimum level possible. Such a scenario seems to be quite challenging.

[3]In what follows, the following terms should be noted:

- In the context of network providers, the notions of primary users and legacy users are used interchangeably.
- Correspondingly, in the context of cognitive radio networks, the notions of secondary users and cognitive radio users are used interchangeably.

- In the *interweave* paradigm, the secondary users must not interfere with the operation of active primary users. Hence, secondary users should be able to accurately detect active primary users and switch bands when it is necessary. This calls for spectrum agility and ability to transmit in different bands.

These three mentioned paradigms can be viewed from another perspective regarding the attitude of primary and secondary users toward each other. For the coexistence of both primary (legacy) users on the one hand and secondary (cognitive radio) users on the other hand in some specific subband in a harmonious manner, there are two spectrum-sharing schemes [57]:

- *Protective-spectrum sharing*, in which legacy owners do not allow the coexistence of secondary users in their nonidle subbands. In other words, when primary users are active and using certain subbands, secondary users should not transmit over those subbands. The interweave paradigm belongs to this category.
- *Aggressive-spectrum sharing*, in which coexistence of primary and secondary users in the same subbands is allowed on the condition that interference power experienced by the primary user's receiver remains below a specified threshold. Both underlay and overlay paradigms belong to this alternative category.

While the protective approach allows secondary users to use only white spaces, the aggressive approach allows them to use gray spaces as well. White spaces refer to subbands, in which only white noise resides. On the other hand, gray spaces refer to subbands that are partially occupied by primary users. Obviously, among the two, aggressive spectrum sharing is less conservative regarding the improvement of spectrum utilization. However, such an approach can only make sense from legacy owners' perspective if it is profitable for them to allow the coexistence of secondary users in their nonidle subbands. In other words, the aggressive approach can only be rationalized if some form of pricing is involved. Market-based models, in which pricing is involved, have been used to study spectrum sharing among one primary user and multiple secondary users as well as spectrum sharing among multiple primary users and one secondary user [58].

In this book, we propose a general framework based on the theory of supply chain networks to study the spectrum market. In the proposed framework, multiple primary users and multiple secondary users as well as multiple brokers are all involved, side by side [59]. Brokers are profit-maximizing entities that buy the right of using spectrum subbands from legacy owners and sell them to secondary users. In a way, they can be viewed as network distributors that are mainly concerned with promotion and sales [60]. In general, brokers are involved in a competitive game among themselves. We may therefore identify two regimes for *dynamic spectrum access* [37]:

- *Open-access regime*: This regime employs open sharing of spectrum among peer secondary users. In the open-spectrum regime, activities of cognitive radio users should not affect the performance of primary users. In other words, the existence of cognitive radio users in a legacy owner's band should not be noticed by the primary users that receive service from that legacy owner. While the legacy owners and their primary users do not need to know anything about the secondary users, secondary users would have to be quite cautious about activities of the primary users. In this regime, secondary users rely on spectrum sensing for identifying spectrum holes [61]. Sweeping the spectrum frequently in search of spectrum holes would be of critical importance for providing seamless secondary communication. In other words, secondary users should always be ready to jump from one subband to another, whenever primary users demand their subbands. Moreover, interference management is performed in a self-organized manner without or with minimum coordination between secondary users in order to maintain an acceptable level of quality of service (QoS) [32].

- *Market-driven regime*: In this alternative regime, pricing is involved and legacy owners gain profit by leasing their idle and partially used subbands to secondary users. In particular, legacy owners are responsible for the probing and management of interference in their own bands. Taking account of maximum allowable interference that guarantees an acceptable level of QoS, legacy users advertise their available subbands for secondary usage. Accordingly, spectrum sensing is not considered as a critical functionality for secondary users' operation in the market-driven regime.

As different as they are, both regimes can be viewed as a spectrum-supply chain network. In addressing them in this manner, they represent two complementary approaches for solving underutilization of the radio spectrum. However, the nature of the games played between the different decision-makers in each one of them is entirely different. Other scenarios can be considered in addition to these two regimes. For instance, *bandwidth exchange* with emphasis on cooperative forwarding is an alternative scenario aimed at enhancing network connectivity and throughput [62]. In this framework, secondary users operate as relays for primary users and in return they get permission to use idle subbands of the corresponding legacy owner for their own data transmission [63].

In each regime, the associated game can be formulated as a *variational inequality (VI) problem*, whose solution is the equilibrium point of the game. The *Nash equilibrium* is commonly used as a standard notion of equilibrium in game theory especially in noncooperative games among selfish players. However, it is not robust against faulty or unexpected behavior, and it is also vulnerable to coalitions [64]. Although the latter sheds light on how to improve the quality of the solution, the former is troublesome. In some games, the equilibrium may not be a valid representation of the outcome of selfish behavior. In other words, instead of converging to an equilibrium point, selfish players may show nonconvergent cyclic behaviors. This may happen even when a unique equilibrium point exists [65]. In a cognitive radio network, spectrum holes can frequently appear or disappear and mobile users can freely start or stop communications. In such a dynamic environment, it is quite likely that we observe fluctuating patterns in users' behaviors [66].

Changes in the number of users as well as the available subbands may occur so fast that users cannot reach an equilibrium point between the time instants that those changes happen. Therefore, compared to the equilibrium point of the game that may not be even reached by the users, the *disequilibrium* behavior of the network, which is governed by its dynamics, more accurately describes the solution quality. Regarding the fact that focusing too much on equilibria, may lead to conclusions about the network behavior, which are qualitatively invalid, a shift of perspective on games from an equilibrium-based analysis toward natural dynamic-process-based description has been suggested [65]. Following this way of thinking, theory of the *projected dynamic systems* is used to derive dynamic models of the spectrum-supply chain for both regimes. These state-space models provide insight on the network disequilibrium behavior over the course of time and facilitate the stability analysis. Such a dynamic-process-based approach distinguishes the current work from other related works such as [67].

1.8 MATHEMATICAL TOOLBOX

A principled basis for the dynamic allocation and management of resources in a cognitive radio network is developed based on the fusion of ideas from game theory, control theory, and optimization.

1.8.1 Game Theory

Game theory provides an analytical toolbox for modeling and analyzing situations in which multiple decision-makers (players) with possibly conflicting interests interact. *Rationality* and *strategically reasoning* are two basic assumptions in game theory. These assumptions reflect that each decision-maker has a well-defined objective and acts based on its knowledge or expectation of other decision-makers' behaviors [68]. Game theory provides a framework for scientifically predicting the future and using this knowledge to engineer it [69].

James Waldegrave discovered the idea of *maxmin* in competitive games and provided the solution for a specific game in 1713. However, it seems that the name "game theory" was first used by Émile Borel in the 1920s. Later, John von Neumann provided proof for the *minimax theorem*. He and Oskar Morgenstern wrote the first book on game theory in 1944 [70]. In addition to discussing noncooperative games, they laid the groundwork for the study of cooperative games by presenting the concept of coalitional games with transferrable utility. A few years later in the beginning of the 1950s, the Nobel Laureate John Nash introduced the key concepts of *Nash Equilibrium* in noncooperative games and *Nash bargaining solution* in cooperative games. In the last few decades, game theory has been a very active field and has benefited from contributions of many researchers such as John Harsanyi and Reinhard Selten, who shared the 1994 Nobel prize in economics with John Nash, and Robert Aumann, who won the Nobel prize in economics in 1995. Harsanyi developed the theory of *Bayesian games*. Selten is well known for his work on *bounded rationality*, which was first proposed by Herbert Simon. Aumann introduced the concept of *correlated equilibrium*.

Coalitional games with nontransferable utilities were also introduced by Aumann and Bezalel Peleg. Although game theory was initially focused on economics, it has been successfully applied for solving problems in other disciplines such as biology. For instance, John Maynard Smith formalized the concept of *evolutionarily stable strategy* (ESS) in the 1970s.

In engineering, in many cases that deal with decentralized control systems, controllers are designed in a centralized manner and then implemented in a decentralized way [71, 72]. This method is not truly decentralized and may cause some problems in practice. Game theory provides a natural framework for analysis and design of truly decentralized control systems. John Nash's paper on "Parallel Control" is perhaps the pioneering work in this area [73]. Influenced by his earlier work on equilibria in noncooperative games [74, 75], Nash proposed to build computers in which components work in a more autonomous way. Başar and Olsder's book on dynamic noncooperative games [76] focuses more on control theoretic aspects and interprets optimal control problems as one-player games. Also, in [77], the robust control problem was interpreted as a zero-sum game in which the controller tries to maximize the system's utility while the environment is trying to minimize the system's utility.

In wireless networks, the radio communication channel is usually shared between different transmitter–receiver (transceiver) pairs. In such environment, multiple users compete for limited resources and the behavior of each user affects the performance of neighboring users. It is therefore not surprising that game theory has attracted the attention of many researchers in the field of communication networks, especially those who are working on cognitive radio.

Recently, several tutorials on game theory have been published for communication engineers. A nice survey on applications of game theory in wired communication systems is presented in [78]. The monograph [79] covers the noncooperative game theory and in the final chapter mentions some research areas in wireless communications and networking that can benefit from game theoretic approaches. The technical report [80] explains the terminology of noncooperative game theory using four simple examples from wireless communications. The concept of equilibria and the related theorems are presented in [81]. The tutorial paper [82] explains the cooperative game theory. In September 2008, *IEEE Journal on Selected Areas in Communications* published a special issue on game theory in communication systems and John Nash wrote a foreword for that. Also, in September 2009, *IEEE Signal Processing Magazine* published another special issue on game theory in signal processing and communications. The latter includes the mentioned tutorial papers on equilibria and cooperative games. Due to the key role that game theory plays in studying cognitive radio networks, the following chapter is dedicated to this topic.

1.8.2 Control Theory

Control engineering is an exciting and challenging field with a multidisciplinary nature and strong mathematical foundation. A control engineer's systematic insight can be easily extended to be utilized in other fields. The present challenge to control engineers is the modeling and control of modern, complex, and interrelated systems.

To face this challenge, we need something dramatically different from traditional control techniques possibly new control structures coming out of the neuroscience world.

Control systems are found throughout nature at the levels of genes, proteins, cells, and entire systems [83]. Some of the natural control systems have unequaled degrees of sophistication [84]. Increased understanding of the scientific and engineering principles behind the living organisms as well as the way they interact with the world and learn from it will lead to fantastic breakthroughs in the design and application of intelligent machines that are truly cognitive.

A living organism interacts with nature through observation and action. Inspired by the perception–action cycle in the brain, a cognitive radio transceiver is built as a closed-loop feedback system, which embodies the radio environment, radio-scene analyzer, feedback channel, and radio-environment actuator. Moreover, a cognitive radio network is a hybrid dynamic system with both continuous and discrete dynamics. Therefore, cognitive radio networks have the potential for presenting a rich spectrum of dynamic behaviors.

1.8.3 Optimization under Uncertainty

In a complex system such as a cognitive radio network, every decision-making process will be a multicriteria optimization problem with possibly conflicting objectives [85]. In order to make certain rational decisions, a user needs to gather information and process it. Data acquisition and computation capabilities of users are limited and they can only make the best decisions regarding their knowledge and resources. Also, real-life cognitive radios are subject to uncertainties that cannot necessarily be dealt with by statistical analysis. In this environment, robust optimization provides an essential tool for making decisions based on the worst-case conditions. According to the Institute of Electrical and Electronics Engineers (IEEE) [86]

> The robustness of a system is the degree to which a system or component can function correctly in the presence of invalid inputs or stressful conditions.

Much too often in the literature, optimality is considered as the driving force for obtaining the best performance possible. Such an objective may well work satisfactorily when considering small-scale applications or toy problems. However, when the application of interest is of a complex or large-scale kind, exemplified by a cognitive radio network, we find ourselves confronted with a much more pressing system requirement: robustness.

Most, if not all, control design strategies exemplified by transmit-power control, are based on the selection of a model for the plant. Selection of the model is influenced by mathematical tractability and prior knowledge that we may have about the plant, a generic term used to describe part of a dynamic system that is supposed to be controlled. Unfortunately, no matter how hard we try and irrespective of all the prior knowledge we may have about the system, there will always be some discrepancy between the actual physical behavior of the plant and the corresponding behavior of the hypothetical model. The response produced at the output of the plant due to a prescribed input signal is determined by the underlying physics of the plant. On the other hand, when the corresponding behavior of the plant is considered, the response of the model due to the same input signal deviates invariably from the actual response of the plant due to unavoidable model uncertainty. The challenge in designing the controller is to make sure that the errors are kept small enough to be acceptable from an operational viewpoint, regardless of all operating conditions that are likely to arise in practice. The following section reviews the dominant sources of uncertainty in cognitive radio networks.

1.9 DOMINANT SOURCES OF UNCERTAINTY IN COGNITIVE RADIO NETWORKS

There are two primary resources in a cognitive radio network: channel bandwidth and transmit power. The operation of the transmit-power controller is complicated by a phenomenon that is peculiar to cognitive radio communication, namely, the fact that spectrum holes come and go, depending on the availability of subbands as permitted by licensed users. To deal with this phenomenon and thereby provide the means for improved utilization of the radio spectrum, a cognitive radio system must have the ability to fill the spectrum holes rapidly and efficiently.

In spectrum sensing that constitutes a basic cognitive function in the receiver, the issue of prime interest is that of variance versus bias of estimation [87]. When we go on to consider the associated cognitive function of transmit-power control in the transmitter, the issue of prime interest is robustness versus optimality [66].

In the context of cognitive radio, the physical plant represents the communication channel between the transmitter and receiver, the radio-scene analyzer plays the role of the sensor, and the radio-environment actuator is the controller. Since the sensor and the actuator are not collocated, they have to be connected by a physical feedback channel and the controller receives the sensor measurements via the feedback channel. Due to the different uncertainty sources in a cognitive radio network, adjusting the transmit power of a cognitive radio requires solving an optimization problem under uncertainty.

The dominant sources of uncertainty in a cognitive radio network are as follows:

- *Primary users*: In a cognitive radio network, spectrum holes come and go, depending on the availability of idle subbands. Therefore, primary users' activities are the cause of *supply-side risk*. Communication patterns of primary users determine the availability and the duration of availability of resources. The availability of the spectrum holes determines the joint feasible set of the resource-allocation optimization problems that are solved by individual secondary users. In other words, it determines the joint set of the action spaces of all secondary users in the corresponding game. As mentioned before, the availability duration of spectrum holes determines the control horizon for the radio-environment actuators of secondary users. Depending on the subbands of interest and the dynamics of activities of primary users in those subbands, two different cases are observed:

a) The activities of the primary users and, therefore, their occupancy of the corresponding subbands are well defined. A good example for this case would be the use of TV bands for cognitive radios.

b) The activities of the primary users and, therefore, the appearance and disappearance of spectrum holes are more dynamic and far less predictable than the former case. A good example for this case would be the use of cellular bands for cognitive radios.

- *Secondary users*: Anytime users can leave the network and new users can join the network in a stochastic manner. This is the cause of *demand-side risk* in the network.
- *Mobility*: Users move all the time. Because of the mobility, the interference that a user causes on other users and mutually the interference that other users cause on that particular user in the network are time-varying.
- *Multiple-time-varying delays*: The feedback channel plays a fundamental role in the design and operation of cognitive radio. Feedback may naturally introduce delay in the control loop and different transmitters may receive statistics of noise and interference with different time delays. Moreover, the sporadic feedback causes users to use outdated statistics to update their power vectors. The time-varying delay in the control loop of each cognitive radio is another source of uncertainty that degrades the performance and may cause stability problems.
- *Noise*: The ambient noise depends on different activities in the environment and is caused by both natural and man-made phenomena.

During the time intervals that the activity of primary users does not change and the available spectrum holes are fixed, two approaches can be taken to deal with the uncertainty caused by joining and leaving of other cognitive radios as well as their mobility: *stochastic optimization* and *robust optimization* [88]. The pros and cons of these two approaches are discussed here.

If there is good knowledge about the probability distribution of the uncertainty sources, then the uncertainty can be dealt with by means of probability and related concepts. In this case, calculation of the expected value will not be an obstacle and, therefore, transmit-power control can be formulated as a stochastic optimization problem.

However, since in practice, little may be known about the probability distribution, the stochastic optimization approach that utilizes the expected value is not a suitable approach. In this case, robust optimization techniques that are based on the worst-case analysis, without involving probability theory, are more appropriate, although such techniques may well be overly conservative in practice. Suboptimality in performance is, in effect, traded in favor of robustness.

Stochastic optimization guarantees some level of performance on average, and sometimes the desired QoS may not be achieved, which means a lack of reliable communication. On the other hand, robust optimization guarantees an acceptable level of performance under the worst-case conditions. It is a conservative approach because real-life systems are not always in their worst behavior, but it can provide seamless communication even in the worst situations. Regarding the dynamic nature of the cognitive radio network and the delay introduced by the feedback channel, the statistics of interference that is used by the transmitter to adjust its power may not represent the current situation of the network. In these cases, robust optimization is equipped to prevent permissible interference power level violation by taking into account the worst-case uncertainty in the interference and noise. Therefore, sacrificing optimality for robustness seems to be a reasonable proposition. However, the use of a predictive model may make it possible for the user to choose the uncertainty set adaptively according to environmental conditions and, therefore, may lead to less conservative designs.

1.10 ISSUE OF TRUSTWORTHINESS

In addition to improving spectral efficiency, cognitive radio networks can lead to socioeconomic benefits due to their impact on competition, innovation, and investment. For instance, they allow for low-speed telemedicine care functions including telenursing and home monitoring through lightweight sparse networks for Internet access. The coalescence of such networks will provide low-cost ubiquitous connectivity and social networking using both fixed network infrastructures and mobile small-scale base stations [89]. Moreover, cognitive radio would be an efficient tool for building *disaster-response networks* [90] or *never-die networks* [91, 92], which should be able to quickly reconnect affected areas by disasters to the rest of the world.

1.11 VISION FOR THE BOOK

The rest of the book is organized as follows:

- Chapter 2 reviews the terminology of game theory with emphasis on noncooperative, cooperative, and minority games, which will be used in design and analysis of cognitive radio networks in the following chapters. The concept of Nash equilibrium is introduced. Reformulation of a noncooperative game as a variational inequality problem is also presented, which paves the way for building dynamic analytic models for networks.
- Chapter 3 discusses the structure of a cognitive radio transceiver and covers its three main functional blocks in detail. These building blocks are radio-scene analyzer, dynamic-spectrum manager, and transmit-power controller. The radio-scene analyzer allows the cognitive radio to perceive the radio ecosystem in order to identify the available spectrum subbands and measure the level of interference plus noise in those subbands. Then, based on the gathered information from the radio ecosystem, a cognitive radio solves the resource-allocation problem in two stages in a hierarchical manner. At the higher level of the

hierarchy, the dynamic spectrum manager selects a set of channels that are more suitable for communications, and then, at the lower level of the hierarchy, the transmit-power controller adjusts the transmit power over the selected channels in a dynamic manner according to the measured levels of interference in those channels. The MTM is discussed in detail as the method of choice for designing the radio-scene analyzer. The notion of cooperative games may be employed to form coalitions among groups of cognitive radios in order to implement a cooperative spectrum sensing mechanism. Two different approaches are suggested for designing the dynamic spectrum manager based on self-organized maps and minority games. These two methods are then compared with each other in terms of performance efficiency regarding different scenarios. The transmit-power controller is designed based on a robust formulation of the iterative waterfilling algorithm in a noncooperative game-theoretic framework.

- Chapter 4 is focused on networks of cognitive radios. It develops a framework based on viewing cognitive radio networks as spectrum-supply chain networks. Two complementary regimes (i.e., open access and market driven) are considered and analytic models for their governing dynamics are derived, which allow for studying both equilibrium and disequilibrium behaviors of the network under study. In the open-access regime, the spectrum-supply chain is a two-tier network consisting of network providers (i.e., legacy owners) and cognitive radio users (i.e., secondary users). On the other hand, in the market-driven regime, the spectrum-supply chain is a three-tier network consisting of network providers, spectrum brokers, and secondary users. In both of the mentioned regimes, a noncooperative game is played among peers in each tier of the network. Theory of variational inequalities is used to derive an equilibrium model for the network, which is the combined outcome of different games played in different tiers of the network. Then, theory of projected dynamic systems is used to derive a state-space model for the network, whose stationary points coincide with the solutions of the corresponding variational inequality (i.e., network equilibrium). This dynamic model allows for analysis of the transient behavior of the network before reaching an equilibrium or transitions between different equilibrium points, when the network is perturbed. Effects of uncertainties and time-varying delays on the behavior of the network are investigated using the analytical dynamic model, and conditions for guaranteeing the network stability are found. Subsequently, theories of evolutionary variational inequalities and projected dynamic systems in Hilbert space are used to extend the developed framework in order to capture the multi-time-scale nature of the network.
- Chapter 5 is dedicated to sustainability of the spectrum-supply chain network. In order to cope with the ever-increasing demand for bandwidth, improving the efficiency of spectrum utilization across licensed and unlicensed bands is of critical importance. For this purpose, an artificial economy is developed based on viewing the licensed and unlicensed bands as private goods and common-pool resources, respectively. The combined outcome of the games played in different tiers of the spectrum-supply chain is a Nash equilibrium, which may

not be Pareto optimal. Moreover, Nash equilibrium is not immune to coalition formation. In order to improve the sustainability of the spectrum-supply chain, the developed artificial economy aims for achieving a Pareto-optimal equilibrium, the so-called Lindahl equilibrium.

- Chapter 6 is focused on cognitive heterogeneous networks (HetNets) with emphasis on economic provisioning for resource sharing. HetNets are viewed as one of the enabling technologies for 5G. An economic model is developed based on decoupling of network infrastructure and spectrum, which facilitates horizontal merger of different networks. In the developed framework, networks merge and split in a dynamic manner to improve their utilities. When networks merge, they may share spectrum, infrastructure, or both depending on the situation in a specific time and location. By the same token, secondary networks may lease spectrum, infrastructure, or both from network providers. In light of merging and splitting of networks, the communication-supply chain network must be optimally designed and redesigned in a dynamic manner.

2

GAME THEORY

This chapter provides a brief account of game theory. Further detail can be found in classic references such as [68] and [76]. The following chapter covers the applications of game theory in cognitive radio networks.

2.1 GAME THEORY TERMINOLOGY

In game theory, the interacting decision-makers are called *players*. In the context of wireless communication systems, the players are users or wireless devices. It is usually assumed that players are *rational* in the sense that they are able to maximize their *utilities*, given all the available information. A player's utility or payoff quantifies the outcome of the game for that player.

However, perfect rationality is questionable. In order to make certain rational decisions, the player needs to gather information and process it. Data acquisition and computation capabilities of players are limited. Therefore, players are usually *bounded rational* in the sense that they can only make the best decisions regarding their knowledge and resources. The players make decisions based on certain decision rules, called *strategies*. The strategy of a player may be a single move or a set of moves. The actual decision for maximizing the utility is called *action* or control.

A game is defined by three basic elements: the set of players, the joint set of the action spaces of all players, which is determined by the constraints on the actions that players can take, and the players' interests, which are quantified as the set of their

Fundamentals of Cognitive Radio, First Edition. Peyman Setoodeh and Simon Haykin.

utilities. A *solution* is a possible outcome of the game. In game theory, reasonable solutions for classes of games are suggested and their characteristics are investigated.

2.1.1 Noncooperative Games versus Cooperative Games

Game theory is divided into two major branches: *noncooperative* games and *cooperative* games. Noncooperative game theory deals with scenarios, in which players compete against each other and selfishly pursue their own interests that are conflicting with other players' interests. Cooperative game theory addresses the formation of cooperative groups of players, called *coalitions*. In each coalition, a group of players coordinate their actions and pool their winnings. These coalitions strengthen the position of the group members in the game.

2.1.2 Static Games versus Dynamic Games

A *static* game is a single-stage game, in which each player acts only once independently of the other players, but this does not correspond to the time slot in general. The game is *dynamic* if the order, in which decisions are made, is important and at least one player uses a strategy that depends on previous actions.

2.1.3 One-Shot Games versus Repeated Games

If a game is played repeatedly, we will have a *repeated game*. In a repeated game, each player will be able to make decisions based on his knowledge about the previous actions of other players. Regarding the role of these information, a repeated game is a dynamic game. Moreover, each player can consider the effect of his current action on future behavior of other players. Therefore, the logic of long-term interaction is inherent in repeated games and the ability of posing threat as well as punishment may lead to emergence of cooperation in these games.

2.1.4 Games with Complete Information versus Games with Incomplete Information

While in *games with complete information*, players are fully informed about each others' strategies and payoffs (but not necessarily the actions), in *games with incomplete information*, at least one player is not aware of strategies and payoffs of at least one other player.

2.1.5 Games with Perfect Information versus Games with Imperfect Information

While in *games with perfect information*, players are fully informed about each others' actions, in *games with imperfect information*, they may be only partially informed and have to make decisions under conditions of uncertainty. They may be [68]

- uncertain about the environmental parameters,
- imperfectly informed about events that happen in the game,
- uncertain about actions of the other players, and
- uncertain about the reasoning of the other players.

As will be explained in the following section, in a wireless network, users have to deal with all of these sources of uncertainty.

It is assumed that players are aware of their alternatives, form expectations about unknowns, have clear preferences, and choose their actions based on solving some optimization problems. In order to model the decision-making process of players, the following basic elements are needed [68]:

- A set of actions from which the decision-maker makes a choice
- A set of possible consequences of these actions
- A consequence function that associates a consequence with each action
- A utility function that specifies player's preferences.

Probability theory and related concepts can be used to model decision-making under uncertainty. If the consequence function is stochastic and known to the decision-maker, a probability distribution on the set of possible consequences can be assigned to each action in the set of actions. Then, the player makes decisions by maximizing his expected utility. In this case, calculation of the expected value will not be an obstacle, when there is a good knowledge about the probability distribution of the uncertain parameters.

However, since in practice, little may be known about the probability distribution, the stochastic optimization approach that utilizes the expected value may not be a suitable approach. In this case, robust optimization techniques that are based on the worst-case analysis, without involving probability theory, are more appropriate, although such techniques may well be overly conservative in practice. Suboptimality in performance is traded in favor of robustness. The formulation of the problem as a robust game is basically a maxmin problem, in which each user tries to maximize its own utility while the environment and the other users are trying to minimize that user's utility [77, 93].

2.2 NONCOOPERATIVE GAMES

A noncooperative game can be represented as a set of coupled optimization problems, in which every player i tries to greedily maximize his payoff u_i.

$$\underset{x_i}{\text{maximize}} \quad u_i(x_i, x_{-i}) \tag{2.1}$$

$$\text{subject to} \quad : \quad x_i \in \mathscr{K}_i, \tag{2.2}$$

where $\mathscr{K}_i$ is the feasible set of player i. Following the notation in the game theory literature, joint actions of the other players are denoted by x_{-i}. In this framework, the idea of a desirable solution is best captured by the notion of *Nash equilibrium*.

2.2.1 Nash Equilibrium

In game theory, the Nash equilibrium is considered to be a concept of fundamental importance. This equilibrium point is a solution such that none of the players have an incentive to deviate from it unilaterally. In other words, in a Nash-equilibrium point, each user's chosen strategy is the best response to the other users' strategies.

2.2.2 Variational Inequalities

A Nash equilibrium game can be reformulated as a variational inequality problem [94]. If $\mathscr{K}_i$ is a closed convex subset of $\mathbb{R}^m$ and u_i is a concave and continuously differentiable function for $i = 1, \dots, n$, then $\mathbf{x}^* = [x_1^*, \dots, x_n^*]^T \in \mathscr{K}$ is a Nash equilibrium of the game if, and only if, it is a solution of the following VI($\mathscr{K}$, $\mathbf{F}$) problem:

$$(\mathbf{x} - \mathbf{x}^*)^T \mathbf{F}(\mathbf{x}^*) \geq 0, \ \forall x \in \mathscr{K}, \tag{2.3}$$

where $\mathscr{K}$ is the joint feasible set of all players and

$$\mathbf{F}(\mathbf{x}) = [\nabla_{x_i} u_i]_{i=1}^n. \tag{2.4}$$

Variational-inequality formulation facilitates adding robustness to the system as well as investigating the existence and uniqueness of the solution. Also, it paves the way for studying the system in a dynamic framework, in which its transient behavior between equilibrium points can be investigated. Further details are provided in Chapter 4. Also, proof of the above statement is provided in Appendix B.

2.3 COOPERATIVE GAMES

Cooperative game theory focuses on what groups of players can achieve. The game is a competition between coalitions of players, rather than between individual players. The worth of a coalition in a game is quantified by the *coalition value*. Therefore, the definition of the coalition value determines the form and type of the game. A group of players can form a coalition, if they can reach a binding agreement about the choice of their actions and the distribution of payoffs. Outcome of a cooperative game is the coalitions that form and the joint action that each coalition can take independently of the remaining players in the game. Although the actions are taken by coalitions, stability of the coalitions depends on individuals' interests because eventually the utilities of the coalitions should be shared among the members of the coalitions based on their individual preferences.

Cooperative coalitional games are divided into two categories: games with *transferable utilities* (TU) and *nontransferable utilities* (NTU). While in coalitional games with TU, the total utility of a coalition can be divided arbitrarily between the coalition members, in coalitional games with NTU, rigid restrictions may exist on the distribution of the utility.

Although the players are eventually interested in maximizing their own utilities, they may try to resolve the conflict by voluntarily taking a course of action that is beneficial for all of them. If there are different options that are more appealing than disagreement for all players, then some form of negotiation will be necessary to reach an agreement, which is good enough for everybody. This behavior can be modeled by Nash bargaining theory.

2.3.1 Nash Bargaining

Nash introduced an axiomatic approach to characterize the properties that a valid bargaining solution should satisfy. The solution must satisfy four properties: linearity, symmetry, Pareto optimality, and independence of irrelevant alternatives. The axiomatic approach does not rely on inter-player comparisons of utilities, and bargaining theory does not model fairness. Instead, it reflects the fact that in the negotiation a stronger player has the upper hand and gets a better share of the resources.

2.4 MINORITY GAMES

In the original form of minority games, a number of players have to choose between two options. Winners are the players that form the minority group. In other words, herding is punished and diversity is rewarded. Players make decisions based on their limited memories of the previous rounds and their information-processing capabilities. Minority games have also been generalized to multichoice and multifrequency cases, in which players make decisions on different timescales [95].

The terminology used in the context of minority games is explained in the following [96]. The collective sum of all players' actions at each round of the game is called *attendance*. Time average variance of the attendance is referred to as *volatility*, which can be regarded as an inverse measure of the efficiency of resource distribution in the game. To be more precise, a high volatility is associated with large fluctuations in attendance, which is a sign of inefficiency of the game. On the other hand, a low volatility is associated with low fluctuations in attendance and shows that the game is fairly efficient.

A payoff function is assigned to strategies, whose value accounts for the predictive ability of a strategy for the next round of the game regardless of players' decisions with respect to choosing or not choosing that particular strategy. If the prediction of a specific strategy about the next round of the game comes out as a winner, that strategy receives a reward and its *cumulated payoff* increases; otherwise, if the prediction comes out as a loser, that strategy is penalized and its cumulated payoff decreases. Payoffs can be updated in two different ways: online and batch. While in the online method, payoffs of different strategies are updated after each round of the game, in the batch method, payoffs are modified only after a fixed number of rounds. Unlike the online method, the batch method is not sensitive to the order of appearance of information about a strategy's prediction in a batch.

If in a minority game, players make decisions and take actions based on the past real history, then, the game is referred to as an *endogenous game*. On the other hand, when players make decisions in a random manner, the game is called an *exogenous game*. In this context, when the batch update method is deployed, the corresponding minority game will look like an exogenous game for the number of rounds that take place between time instances that payoffs are update.

Regarding the dynamics of the game, two different phases are distinguished: symmetric and asymmetric. The symmetric phase is also referred to as crowded or unpredictable phase, and the asymmetric phase is also known as uncrowded, dilute, or predictable phase. Transition between these two phases is an important issue in studying the system behavior at the macroscopic level. These phases are characterized by an order parameter known as *predictability*. It is a measure of the probabilistic outcome of the attendance, when players have access to a certain amount of information. In this regard, the ratio of the total amount of possible information to the population size is an important parameter that controls the phase transition. In other words, this parameter has a critical value for which a phase transition occurs. It can also be used to rescale the macroscopic observables of the game for different amounts of information and population size.

2.5 CONCLUDING REMARKS

This chapter covered a brief account of game theory to provide the necessary background for the following chapters. The next chapter is dedicated to different building blocks of a cognitive radio transceiver including radio-scene analyzer, dynamic spectrum manager, and transmit power controller. As will be explained later, these three building blocks can be designed regarding different game-theoretic frameworks. To be more precise, a radio-scene analysis mechanism can be implemented based on a cooperative game for spectrum sensing, dynamic spectrum management can benefit from minority games to encourage radios to choose channels with lower levels of interference, and noncooperative games provide the required framework for transmit power control. When it comes to networks of cognitive radios, network dynamics are governed by the outcome of different games that are played at different layers of the network. This issue will be extensively studied in Chapter 4. Existence and uniqueness of an equilibrium for the network may be guaranteed through repeated games for packet forwarding and relaying.

3

COGNITIVE RADIO TRANSCEIVER

This chapter covers the four basic building blocks of a cognitive radio transceiver: spectrum sensing, feedback channel, dynamic spectrum management, and transmit power control. Through spectrum sensing, the cognitive radio transceiver perceives the radio ecosystem to extract the required information for maintaining a seamless communication. The spectrum-sensing process must provide the cognitive radio with the information about available spectrum subbands and the level of interference in those subbands. Then, dynamic spectrum manager selects a set of channels from the available subbands for communications. Subsequently, transmit power controller adjusts the transmit power over the selected channels according to the interference level in those channels. Information about the state of the selected forward channels is sent to the transmitter from the receiver through the feedback channel.

This chapter provides extensive guidelines on designing the spectrum-sensing mechanism, dynamic spectrum manager, and transmit power controller. The multi-taper method (MTM) is proposed as the method of choice for spectrum sensing. For dynamic spectrum management, two methods are discussed and compared against each other: one is based on self-organized maps (SOMs) and the other one is based on minority games. The transmit power controller is designed based on a robust version of the iterative waterfilling algorithm (IWFA). After covering the main building blocks, the information value and its flow in the perception–action cycle is discussed briefly.

Fundamentals of Cognitive Radio, First Edition. Peyman Setoodeh and Simon Haykin.

3.1 SPECTRUM SENSING

With spectrum holes playing a critical role in the underlying theory and design of cognitive radio networks, the key question is as follows: Where are they to be found in the radio spectrum? To set the stage for addressing this basic question, it is instructive that we reaffirm two kinds of spectrum holes:

1. *White spaces*, which are free of RF interferers, except for *noise* due to natural and/ or artificial sources
2. *Gray spaces*, which are partially occupied by interferers and noise.

The first place where we may find spectrum holes is the *television (TV) band*. The transition of all terrestrial television broadcasting from analog to digital, using the Advanced Television Systems Committee (ATSC) Standard, was accomplished in 2009 in North America. Moreover, in November 2008, the FCC in the United States ruled that access to the *ATSC-digital television (DTV)* band be permitted for wireless devices. Thus, for the first time ever, the way was opened in 2009 for the creation of "white spaces" for use by low-power cognitive radios. The availability of these white spaces will naturally vary across time and from one geographic location to another. In reality, however, noise is not likely to be the sole occupant of the ATSC-DTV band when a TV broadcasting station is switched off. Rather, interfering signals of widely varying power levels do exist below the DTV pilot. In other words, some of the subbands constituting the ATSC-DTV band may indeed be actually "gray," not "white."

The most common natural source of noise encountered at the front end of communication receivers is *thermal noise*, which is justifiably modeled as additive white Gaussian noise (AWGN). By far, the most important artificial source of noise in mobile communications is *man-made noise*, which is radiated by different kinds of electrical equipment across a frequency band extending from about 2 MHz to about 500 MHz [97]. Unlike thermal noise, man-made noise is *impulsive* in nature; hence the reference to it as *impulsive noise*. In urban areas, the impulsive noise generated by motor vehicles is a major source of interference to mobile communications. With the statistics of impulsive noise being radically different from the Gaussian characterization of thermal noise, the modeling of noise in a white space due to the combined presence of Gaussian noise and impulsive noise in urban areas may complicate procedures for identifying spectrum holes.

Consider next the commercial cellular networks deployed all over the world. In the current licensing regime, only primary users have exclusive rights to transmit. However, it is highly likely to find small spatial footprints in large cells where there are no primary users. Currently, opportunistic low-power usage of the cellular spectrum is not allowed in these areas, even though such usage by cognitive radios in a *femtocell* with a small base station is not detrimental to the primary user [98]. Thus, spectrum holes may also be found in commercial cellular bands; naturally, spread of the spectrum holes varies over time and space. In any event, account has to be taken of interference arising from conflict relationships between transmitters of various radio infrastructure providers that coexist in a region [99]. Consequently, the spectrum holes found in cellular bands may also not all be white spaces.

The important point to take from this discussion is that, regardless of where the spectrum holes exist, be they in the ATSC-DTV band or cellular band, we are confronted with the practical reality that the spectrum holes may be made up of white and gray spaces. This possibility may, therefore, complicate applicability of a simple hypothesis-testing procedure that designates each subband as black (blocked space) or white (exploitable) space, using energy detection or cyclostationarity characterization.

One other comment is in order regarding a study conducted in Aachen, Germany, which is not listed in Table 1.1. Therein, measurements over a period of 7 days, next to the main railway station in the band 20–3000 MHz, showed that spectrum utilization was 32% for the indoor environment and about 100% for the outdoor environment. However, in such a place, the sensors were exposed to high-level ambient noise, and the inability of energy detectors to distinguish man-made noise from primary users' signals led to this unexpectedly high occupancy measurement [100]; it is for reasons just mentioned, this particular study is not listed in Table 1.1. The important lesson to learn from this study is that we should be cautious about the spectrum-sensing method that we adopt in order to avoid such misleading results. The spectrum-sensing method, which is explained in this section, is built on the technique presented in [87].

3.1.1 Attributes of Reliable Spectrum Sensing

In light of these practical realities discussed above, we may now identify the desirable attributes of a spectrum sensor for this should be cognitive radio applications:

1. *Detection of spectrum holes* and their *reliable* classification into white and gray spaces; this classification may require an *accurate estimation of the power spectrum*, particularly when the spectrum hole is of a gray-space kind.
2. *Accurate spectral resolution* of spectrum holes, which is needed for efficient utilization of the radio spectrum; after all, this efficient utilization is the driving force behind cognitive radio.
3. *Estimation of direction-of-arrival (DOA) of interferers*, which provides the cognitive radio a sense of *spatial direction*.
4. *Time-frequency analysis* (TFA) for highlighting *cyclostationarity*, which could be used as an additional method for the reinforcement of spectrum-hole detection and also *modulated-signal classification*, when the subband of interest is occupied by a primary user.

3.1.2 The Multitaper Method

A radio-sensing scheme that satisfies the desirable attributes of spectrum sensing, just described, is the MTM, also known as the multiple window method [101]. The MTM is a spectrum estimation procedure that is based on *Slepian sequences*, a unique characteristic of which is that of *maximizing the input signal energy in the smallest spectral bandwidth possible* [102]. In other words, viewing the Slepian sequence as a window, it is the optimal window superior to any other signal processing window employed in a spectral estimation (e.g., Hamming and Hanning windows). With optimum spectrum utilization as the design criterion of interest, the MTM may therefore be viewed as the "best" candidate for spectrum sensing.

We say so for the following reasons, over and above the above-mentioned desirable attributes [13]:

1. In multitaper spectral estimation, the bias is decomposed into two quantifiable components:
   - *local bias*, due to frequency components residing inside the user-selectable band from $f - W$ to $f + W$, and
   - *broadband bias*, due to frequency components found outside this band.
2. The *resolution* of a multitaper spectral estimator is naturally defined by the bandwidth of the passband, namely $2W$.
3. Multitaper spectral estimators offer an easy-to-quantify *trade-off between bias and variance*; accordingly, the bias-variance dilemma is replaced via the variance-resolution dilemma.
4. Direct spectrum estimation can be performed with more than just two *degrees of freedom* (DoFs); typically, the DoFs vary from 6 to 10, depending on the time-bandwidth product used in the estimation.
5. Multitaper spectral estimation has a built-in form of regularization; in other words, multitaper spectral estimation provides an analytic basis for computing the best approximation to a desired power spectrum, which is not possible from observable data alone.
6. Through the inclusion of a multiple array of antennas in the receiver, the MTM acquires a space-time processing capability. Hence, information on the state of the radio environment (i.e., spectrum holes) in time as well as space can be computed. This kind of information can be of practical importance to the cognitive process of attention in cognitive radio.
7. The MTM provides a rigorous mathematical basis for computing the cyclostationary characteristic of incoming radio signals [87], which can be exploited for identifying the legacy user responsible for occupying a subband of the radio spectrum of specific interest; as such, cyclostationarity can provide another way of further enhancing the detection of spectrum holes.
8. Last, but by no means least, multitaper spectral estimates can be used to distinguish spectral line components within the band $(f - W, f + W)$ by including the *harmonic F-test*.

Furthermore, from a computational perspective, by precomputing the Slepian sequences and using the state-of-the-art fast Fourier transform (FFT) algorithm, namely the so-called *fastest Fourier transform in the west* (FFTW) [103], computation of the MTM for spectrum sensing can be accomplished in a matter of 5–20 μs.

Even faster spectrum-sensing time is expected to be achieved in the future as the speed and computing power of advanced mobile-device processors are continually improved. In [87], experimental results are presented using real-life data, which clearly demonstrate the practical effectiveness of the MTM for spectrum sensing.

Let t denote *discrete time*. Let the time series $\{x(t)\}_{t=0}^{N-1}$ represent the *baseband* version of the received RF signal with respect to the center frequency of the RF band under scrutiny; the term baseband means that the center frequency of the signal is moved (demodulated) down to 0 Hz. Given this time series, the MTM determines the following parameters [101]:

- an orthonormal sequence of *Slepian tapers*, denoted by $\left\{v_t^{(k)}\right\}_{t=0}^{N-1}$; and
- a corresponding set of Fourier transforms

$$X_k(f) = \sum_{t=0}^{N-1} x(t)v_t^{(k)} \exp(-j2\pi ft), \tag{3.1}$$

where $k = 0, 1, \ldots, K-1$. The energy distributions of the *eigenspectra*, defined by $|X_k(f)|^2$ for varying k, are concentrated inside a *resolution bandwidth* $2W$. The *time-bandwidth product*

$$C_o = NW$$

bounds the number of tapers (windows) as shown by

$$K \leq \lfloor 2NW \rfloor \tag{3.2}$$

which, in turn, defines the *DoF* available for controlling the variance of the multitaper spectral estimator. The choice of parameters C_o and K provides a trade-off between spectral resolution, bias, and variance. The bias of these estimates is largely controlled by the largest eigenvalue, denoted by $\lambda_0(N, W)$, which is given asymptotically by Thomson [101]

$$1 - \lambda_0 \approx 4\pi\sqrt{C_o}\exp(-2\pi C_o)$$

This formula gives the fraction of the *total sidelobe energy*, that is, the total leakage into frequencies outside the interval $(-W, W)$; the total sidelobe energy decreases very rapidly with C_o as can be seen in Table 3.1. A natural spectral estimate, based on the first few eigenspectra that exhibit the least sidelobe leakage, is given by [101, 104, 105]

$$\hat{S}(f) = \frac{\sum_{k=0}^{K-1} \lambda_k(f)|X_k(f)|^2}{\sum_{k=0}^{K-1} \lambda_k(f)}, \tag{3.3}$$

where $X_k(f)$ is the Fourier transform defined in (3.1) and $\lambda_k(f)$ is the eigenvalue associated with the kth eigenspectrum. The denominator in (3.3) makes the estimator $\hat{S}(f)$ unbiased.

TABLE 3.1 Leakage Properties of the Lowest-Order Slepian Sequence as a Function of the Time-Bandwidth Product C_o (Column 1)

$C_o = NW$	$1 - \lambda_0$	dB
4	3.05×10^{-10}	−95
6	1.31×10^{-15}	−149
8	5.26×10^{-21}	−203
10	2.05×10^{-26}	−257

Column 2 of the table gives the asymptotic value of $1 - \lambda_0(C_o)$, and Column 3 is the same (total sidelobe energy) expressed in decibels (relative to total energy in the signal).

The multitaper spectral estimator of (3.3) is intuitively appealing in the way it works: As the number of tapers, K, increases, the eigenvalues decrease, causing the eigenspectra to be more contaminated by leakage. However, the eigenvalues themselves counteract this effect by reducing the weighting applied to higher-leakage eigenspectra.

While the lower-order eigenspectra have excellent bias properties, there is some degradation as the order K increases toward the limiting value defined in (3.2). In [101], a set of *adaptive weights*, denoted by $\{d_k(f)\}$, is introduced to downweight the higher-order eigenspectra. Using a mean-square error optimization procedure, the following formula for the weights is derived:

$$d_k(f) = \frac{\sqrt{\lambda_k(f)}S(f)}{\lambda_k(f)S(f) + \mathbf{E}[B_k(f)]}; \quad k = 0, 1, \ldots, K-1, \tag{3.4}$$

where $S(f)$ is the true power spectrum, $B_k(f)$ is the broadband bias of the kth eigenspectrum, and $\mathbf{E}$ is the statistical expectation operator. Moreover, we find that

$$\mathbf{E}[B_k(f)] \leq (1 - \lambda_k(f))\sigma^2; \quad k = 0, 1, \ldots, K-1, \tag{3.5}$$

where σ^2 is the *process variance*, defined in terms of the time series $x(t)$ by

$$\sigma^2 = \frac{1}{N}\sum_{t=0}^{N-1} |x(t)|^2. \tag{3.6}$$

In order to compute the adaptive weights $d_k(f)$ using (3.4), we need to know the true spectrum $S(f)$. Clearly, if we know $S(f)$, then there would be no need to perform any spectrum estimation in the first place. Nevertheless, the formula of (3.4) is useful in setting up an *iterative procedure for computing the adaptive spectral estimator*, as shown by

$$\hat{S}(f) = \frac{\sum_{k=0}^{K-1} |d_k(f)|^2 \hat{S}_k(f)}{\sum_{k=0}^{K-1} |d_k(f)|^2}, \tag{3.7}$$

where

$$\hat{S}_k(f) = |X_k(f)|^2; \quad k = 0, 1, \ldots, K-1. \tag{3.8}$$

Note that if we set $|d_k(f)|^2 = \lambda_k(f)$ for all k, then the estimator of (3.7) reduces to that of (3.3).

Next, setting $S(f)$ equal to the spectrum estimate $\hat{S}(f)$ in (3.4), then substituting the new equation into (3.7) and collecting terms, we get (after simplifications)

$$\sum_{k=0}^{K-1} \frac{\lambda_k(f)(\hat{S}(f) - \hat{S}_k(f))}{(\lambda_k(f)\hat{S}(f) + \hat{B}_k(f))^2} = 0, \tag{3.9}$$

where $\hat{B}_k(f)$ is an estimate of the expectation $\mathbf{E}[B_k(f)]$. Using the upper bound of (3.5), we may set

$$\hat{B}_k(f) = (1 - \lambda_k(f))\sigma^2; \quad k = 0, 1, \ldots, K-1. \tag{3.10}$$

We now have all that we need to solve for the null condition of (3.9) via the *recursion*

$$\hat{S}^{j+1}(f) = \left[\sum_{k=0}^{K-1} \frac{\lambda_k(f)\hat{S}_k^{(j)}(f)}{(\lambda_k(f)\hat{S}^{(j)}(f) + \hat{B}_k(f))^2}\right] \times \left[\sum_{k=0}^{K-1} \frac{\lambda_k(f)}{(\lambda_k(f)\hat{S}^{(j)}(f) + \hat{B}_k(f))^2}\right]^{-1}, \tag{3.11}$$

where j denotes an iteration step, that is, $j = 0, 1, 2, \ldots$. To initialize the recursion of (3.11), we may set $\hat{S}^j(0)$ equal to the average of the two lowest-order eigenspectra. Convergence of the recursion is usually rapid, with successive spectral estimates differing by less than 5% in 5–20 iterations [105]. The result obtained from (3.11) is substituted into (3.4) to obtain the desired weights, $d_k(f)$.

A useful by-product of the adaptive spectral estimation procedure is a *stability measure of the estimates*, given by

$$v(f) = 2\sum_{k=0}^{K-1} |d_k(f)|^2, \tag{3.12}$$

which is the approximate DoF for the estimator $\hat{S}(f)$, expressed as a function of frequency f. If $\overline{v}$, denoting the average of $v(f)$ over frequency f, is significantly less than $2K$, then the result is an indication that either the bandwidth $2W$ is too small, or additional *prewhitening* of the time series $x(t)$ should be used.

The importance of prewhitening cannot be stressed enough for RF data. In essence, prewhitening reduces the dynamic range of the spectrum by filtering the data, prior to processing. The resulting residual spectrum is nearly flat or "white." In particular, leakage from strong components is reduced, so that the fine structure of weaker components is more likely to be resolved [105].

(i) Estimation of the power spectrum based on the formula of (3.3) is said to be *incoherent*, because the kth eigenspectrum $|X_k(f)|^2$ ignores phase information for all values of the index k.

(ii) For the parameters needed to compute the multitaper spectral estimator (3.3), recommended values (within each data section) are as follows:

- Parameter $C_o = 4$, possibly extending up to 10
- Number of Slepian tapers: $K = 10$, possibly extending up to 16.

These values are needed, especially when the dynamic range of the RF data is large.

(iii) If, and when, the number of tapers is increased toward the limiting value $2NW$, then the adaptive multitaper spectral estimator should be used.

3.1.3 Space-Time Processing

As already discussed, the MTM is theorized to provide a reliable and accurate method of estimating the power spectrum of RF stimuli as a function of frequency. As such, in the MTM we have a desirable method for identifying spectrum holes and estimating their average-power contents. In analyzing the radio scene in the local neighborhood of a cognitive radio receiver, however, we also need to have a *sense of direction*, so that the cognitive radio is able to *listen* to incoming interfering signals from unknown directions. What we are signifying here is the need for *space-time processing*. To this end, we may *employ a set of sensors to properly "sniff" the RF environment along different directions.*

To elaborate on this matter, consider an array of M antennas sensing the environment. For the kth Slepian taper, let $X_k^{(m)}(f)$ denote the complex-valued Fourier transform of the input signal $x(t)$ computed by the mth sensor in accordance with (3.1), and $m = 0, 1, \ldots, M-1$. With $k = 0, 1, \ldots, K-1$, we may then construct the M-by-K *spatiotemporal complex-valued matrix*

$$\mathbf{A}(f) = \begin{bmatrix} a_0^{(0)}X_0^{(0)} & a_1^{(0)}X_1^{(0)} & \cdots & a_{K-1}^{(0)}X_{K-1}^{(0)} \\ a_0^{(1)}X_0^{(1)} & a_1^{(1)}X_1^{(1)} & \cdots & a_{K-1}^{(1)}X_{K-1}^{(1)} \\ & & \vdots & \\ a_0^{(M-1)}X_0^{(M-1)} & a_1^{(M-1)}X_1^{(M-1)} & \cdots & a_{K-1}^{(M-1)}X_{K-1}^{(M-1)} \end{bmatrix}, \tag{3.13}$$

where each row of the matrix is produced by RF stimuli sensed at a different gridpoint, each column is computed using a different Slepian taper, and $a_k^{(m)}$ represents coefficients accounting for different areas of the gridpoints.

To proceed further, we make two necessary assumptions:

1. The number of Slepian tapers, K, is larger than the number of sensors M; this requirement is needed to avoid "spatial undersampling" of the RF environment.
2. Except for being synchronously sampled, the M sensors operate independently of each other; this second requirement is needed to ensure that the rank of the matrix $\mathbf{A}(f)$ (i.e., the number of linearly independent rows) is equal to M.

In physical terms, each entry in the matrix $\mathbf{A}(f)$ is produced by two contributions, one due to additive ambient noise at the front end of the sensor and the other due to the incoming RF stimuli. Insofar as spectrum sensing is concerned, the primary contribution of interest is that due to RF stimuli. In this context, an effective tool for denoising is *singular value decomposition* (SVD).

The SVD is a generalization of *principal-components analysis*, or *eigen-decomposition.* While eigen-decomposition involves a single orthonormal matrix, the SVD involves a pair of orthonormal matrices, which we denote by an M-by-M matrix $\mathbf{U}$ and a K-by-K matrix $\mathbf{V}$. Thus, applying the SVD to the spatiotemporal matrix $\mathbf{A}(f)$, we may express the resulting decomposition as follows [106]:

$$\mathbf{U}^{\dagger}\mathbf{A}(f)\mathbf{V}(f) = \begin{bmatrix} \Sigma(f) \\ \text{----} \\ \mathbf{0} \end{bmatrix}, \tag{3.14}$$

where the superscript $\dagger$ denotes *Hermitian transposition* and $\sum(f)$ is an M-by-M diagonal matrix, the kth element of which is denoted by $\sigma_k(f)$. Figure 3.1 shows an insightful depiction of this decomposition; to simplify the depiction, dependence on the frequency f has been ignored.

Henceforth, the system described by the spatiotemporal matrix $\mathbf{A}(f)$ of (3.13), involving K Slepian tapers, M sensors, and decomposition of the matrix in (3.14), is referred to as the *MTM-SVD processor*. Note that with the spatiotemporal matrix $\mathbf{A}(f)$ being frequency dependent, and likewise for the unitary matrices $\mathbf{U}(f)$ and $\mathbf{V}(f)$, the MTM-SVD processor is actually performing *tensor analysis*.

To understand the underlying signal operations embodied in the MTM-SVD processor, we begin by reminding ourselves of the orthonormal properties of matrices $\mathbf{U}$ and $\mathbf{V}$ that hold for all f, as shown by

$$\mathbf{U}(f)\mathbf{U}^{\dagger}(f) = \mathbf{I}_M$$

and

$$\mathbf{V}(f)\mathbf{V}^{\dagger}(f) = \mathbf{I}_K,$$

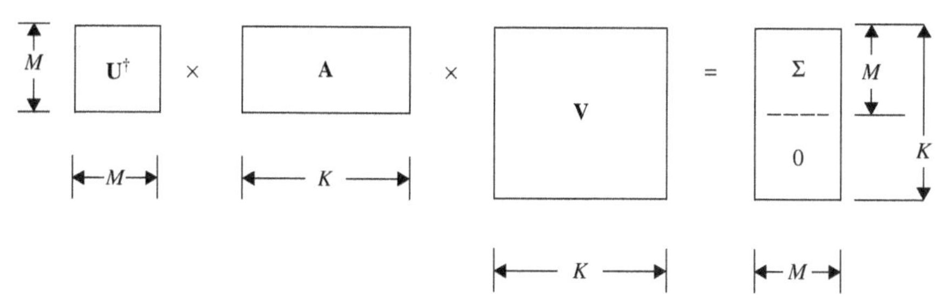

FIGURE 3.1 Diagrammatic depiction of singular value decomposition applied to the matrix $\mathbf{A}$ of (3.13). Source: Haykin (2009) [48]. Reproduced with the permission of IEEE.

where $\mathbf{I}_K$ and $\mathbf{I}_M$ are K-by-K and M-by-M identity matrices, respectively. Using this pair of relations in (3.14), we obtain the following decomposition of the matrix $\mathbf{A}(f)$ (after some straightforward manipulations):

$$\mathbf{A}(f) = \sum_{m=0}^{M-1} \sigma_m \mathbf{u}_m \mathbf{v}_m^{\dagger}. \tag{3.15}$$

The $\sigma_m(f)$ is called the mth *singular value* of the matrix $\mathbf{A}(f)$, $\mathbf{u}_m(f)$ is called the *left-singular vector*, and $\mathbf{v}_m(f)$ is called the *right-singular vector*. In analogy with *principal-components analysis*, the decomposition of (3.15) may be viewed as one of the *principal modulations* produced by the incoming RF stimuli [30, 107]. According to this decomposition, the singular value $\sigma_m(f)$ scales the mth principal modulation computed by the MTM-SVD processor.

The M singular values, constituting the diagonal matrix $\sum(f)$ in (3.14), are all real numbers. The higher-order singular values, namely, $\sigma_M(f), \ldots, \sigma_{K-1}(f)$, are all zero; they constitute the null matrix $\mathbf{0}$ in (3.14).

Using (3.15) to form the matrix product $\mathbf{A}(f)\mathbf{A}^{\dagger}(f)$, and invoking the orthonormal property of the unitary matrix $\mathbf{V}(f)$, we have the eigen-decomposition

$$\mathbf{A}(f)\mathbf{A}^{\dagger}(f) = \sum_{m=0}^{M-1} \sigma_m^2(f)\mathbf{u}_m(f)\mathbf{u}_m^{\dagger}(f),$$

where $\sigma_M^2(f)$ is the mth eigenvalue of the eigen-decomposition. Similarly, forming the other matrix product $\mathbf{A}^{\dagger}(f)\mathbf{A}(f)$ and invoking the orthonormal property of the unitary matrix $\mathbf{U}(f)$, we have the alternative eigen-decomposition

$$\mathbf{A}^{\dagger}(f)\mathbf{A}(f) = \sum_{k=0}^{M-1} \sigma_k^2(f)\mathbf{v}_k(f)\mathbf{v}_k^{\dagger}(f),$$

where the eigenvalues for $k = M, \ldots, K-1$ are all zero.

Recalling that the index m signifies a sensor and the index k signifies a Slepian taper, we may now make three statements on the multiple operations being performed by the MTM-SVD processor:

1. The mth eigenvalue $\sigma_m^2(f)$ is defined by

$$\sigma_m^2(f) = \sum_{k=0}^{K-1} |a_k^{(m)}(f)|^2 |X_k^{(m)}(f)|^2.$$

Setting $|a_k^{(m)}(f)|^2 = \lambda_k^{(m)}(f)$ and dividing $\sigma_m^2(f)$ by $\sum_{k=0}^{K-1} \lambda_k^{(m)}(f)$, we get

$$\hat{S}^{(m)}(f) = \frac{\sum_{k=0}^{K-1} \lambda_k^{(m)}(f)|X_k^{(m)}(f)|^2}{\sum_{k=0}^{K-1} \lambda_k^{(m)}(f)}; \tag{3.16}$$
$$m = 0, 1, \ldots, M-1,$$

which is a rewrite of the formula in (3.3), specialized for sensor m. We may therefore make the statement:

> The eigenvalue $\sigma_m^2(f)$, except for the scaling factor $\sum_{k=0}^{K-1} \lambda_k^{(m)}(f)$, provides the desired multitaper spectral estimate of the incoming interfering signal picked up by the mth sensor.

2. Since the index m refers to the mth sensor, we make the second statement:

> The left singular vector $\mathbf{u}_m(f)$ defines the direction of the interfering signal picked up by the mth sensor at frequency f.

3. The index k refers to the kth Slepian taper; moreover, since $\sigma_k^2(f) = \sigma_m^2(f)$ for $k = 0, 1, \ldots, M-1$, we may make the third and last statement:

> The right singular vector $\mathbf{v}_m(f)$ defines the multitaper coefficients for the mth interferer's waveform.

Most importantly, with no statistical assumptions on the additive ambient noise in each sensor or the incoming RF interferers, we may go on to state that the nonparametric MTM-SVD processor is indeed *robust*.

The enhanced signal-processing capability of the MTM-SVD processor just described is achieved at the expense of increased computational complexity. To elaborate, with N data points and signal bandwidth $2W$, there are N different frequencies with spectral resolution $2W/N$ to be considered. Accordingly, the MTM-SVD processor has to perform a total of N SVDs on matrix $\mathbf{A}(f)$. Note, however, the size of the wavenumber spectrum (i.e., the spatial distribution of the interferers) is determined by the number of sensors, M, which is considerably smaller than the number of data points, N. Most importantly, the wavenumber is computed in *parallel*. With the computation being performed at each frequency f, each of the M sensors sees the full spectral footprint of the interferer pointing along its own direction; the footprint is made up of N frequency points with a spectral resolution of $2W/N$.

Summing up, the MTM-SVD processor has the capability to sense the surrounding RF environment both in frequency as well as space, the resolutions of which are, respectively, determined by the number of data points and the number of sensors deployed.

3.1.4 Time-Frequency Analysis

The MTM-SVD processor rests its signal-processing capability on two dimensions of sensing:

- *Frequency*, which is necessary for identifying the location of spectrum holes along the frequency axis

- *Space*, which provides the means for estimating wavenumber spectra of the RF environments.

However, for a cognitive radio to be fully equipped to sense its local neighborhood, there is a third dimension of sensing that is just as important: *time*. The inclusion of time is needed for the cognitive radio receiver to sense the type of modulation employed by the primary user, for example, so as to provide for a harmonious relationship with the primary user if, and when, it is needed. This need calls for *TFA*.

The statistical analysis of nonstationary signals has had a rather mixed history. Although the general second-order theory was published during 1946 by Loève [108], it has not been applied nearly as extensively as the theory of stationary processes published only slightly earlier by Wiener and Kolmogorov. There were, at least, four distinct reasons for this neglect, as summarized in [109]:

(i) Loève's theory was *probabilistic*, not statistical, and there does not appear to have been successful attempts to find a statistical version of the theory until sometime later.

(ii) At that time of publications, over six decades ago, the mathematical training of most engineers and physicists in signals and stochastic processes was minimal and, recalling that even Wiener's delightful book was referred to as "The Yellow Peril" because of the color of its cover, it is easy to imagine the reception that a general nonstationary theory would have received.

(iii) Even if the theory had been commonly understood at the time, and good statistical estimation procedures had been available, the computational burden would probably have been overwhelming. This was the era when Blackman–Tukey estimates of the stationary spectrum were developed, not because they were great estimates but, primarily, because they were simple to understand in mathematical terms and, before the (re)invention of the FFT algorithm, computationally more efficient than other forms.

(iv) Finally, it cannot be denied that the Loève theory of nonstationary processes was harder to grasp than that of stationary processes.

In any event, confronted with the notoriously unreliable nature of a wireless channel, we have to find some way to account for the nonstationary behavior of a signal at the channel output, and therefore *time* (implicitly or explicitly), in a description of the signal picked up by the receiver. Given the desirability of working in the frequency domain for well-established reasons, we may include the effect of time by adopting a time-frequency description of the signal. During the last three decades, many papers have been published on various estimates of time-frequency distributions; see, for example, Cohen's book [110] and the references therein. In most of this work, the signal is assumed to be *deterministic*. In addition, many of the proposed estimators are constrained to match time and frequency *marginal* density conditions. For a continuous-time signal $x(t)$, the *time marginal* is required to satisfy the condition

$$\int_{-\infty}^{\infty} D(t,f)\mathrm{d}f = |x(t)|^2, \tag{3.17}$$

where $D(t,f)$ is the *time-frequency distribution* of the signal. Similarly, if $X(f)$ is the Fourier transform of $x(t)$, the *frequency marginal* must satisfy the second condition

$$\int_{-\infty}^{\infty} D(t,f)\mathrm{d}t = |X(f)|^2. \tag{3.18}$$

Given the large differences observed between waveforms collected on sensors spaced short distances apart, the time marginal requirement is a rather strange assumption. Worse, the frequency marginal is, except for a factor of $1/N$, just the periodogram of the signal. It has been known, well before the first periodogram was computed [111], that the periodogram is badly biased and inconsistent. An *inconsistent estimate* is one where the variance of the estimate does not decrease with sample size. Rayleigh did not use the term "inconsistent" because it was not introduced as a statistical term until Fisher's famous paper nearly 30 years later. Thus, we do not consider matching marginal distributions, as commonly defined in the literature, to be important.

Nonstationarity is an inherent characteristic of most, if not all, of the stochastic processes encountered in practice. Yet, despite its highly pervasive nature and practical importance, not enough attention is paid in the literature to the characterization of nonstationary processes in a mathematically satisfactory manner.

To this end, consider a complex continuous stochastic process, a sample function of which is denoted by $x(t)$, where t denotes *continuous time*. We assume that the process is *harmonizable* [108], so that it permits the *Cramér representation*

$$x(t) = \int_{-1/2}^{1/2} \exp(j2\pi\nu t)\mathrm{d}Z_x(\nu), \tag{3.19}$$

where $\mathrm{d}Z_x(\nu)$ is an *orthogonal increment process* associated with $x(t)$; the dummy variable ν has the same dimension as frequency. The bandwidth of $x(t)$ has been normalized to *unity* for convenience of presentation; consequently, as indicated in (3.19), the integration extends with respect to ν over the interval $[-1/2, +1/2]$. As before, it is assumed that the processor has zero mean, that is, $\mathbf{E}[x(t)] = 0$ for all time t; correspondingly, we have $\mathbf{E}[Z_x(\nu)] = 0$ for all ν. (The formula of (3.19) is also the starting point in formulating the MTM.)

To set the stage for introducing the statistical parameters of interest, we define the covariance function

$$\begin{aligned}\Gamma_L(t_1, t_2) &= \mathbf{E}\{x(t_1)x^*(t_2)\} \\ &= \int_{-\infty}^{\infty}\int_{-\infty}^{\infty} \exp(j2\pi(t_1 f_1 - t_2 f_2))\gamma_L(f_1, f_2)\mathrm{d}f_1\mathrm{d}f_2,\end{aligned} \tag{3.20}$$

where, in the first line, the asterisk denotes complex conjugation. Hereafter, the generalized two-frequency spectrum $\gamma_L(f_1, f_2)$ in the integrand of the second line in (3.20) is referred to as the *Loève spectrum*. With $X(f)$ denoting the Fourier transform of $x(t)$, the Loève spectrum is formally defined as follows:

$$\gamma_L(f_1, f_2)\mathrm{d}f_1\mathrm{d}f_2 = \mathbf{E}[\mathrm{d}Z_x(f_1)\mathrm{d}Z_x^*(f_2)], \tag{3.21}$$

where, as before, $dZ_x(f)$ is an orthogonal increment associated with $x(t)$. Care should be exercised in distinguishing the Loève spectrum $\gamma_L(f_1,f_2)$ from the bispectrum $B(f_1,f_2)$. Both are functions of two frequencies, but the Loève spectrum $\gamma_L(f_1,f_2)$ is a *second*-moment description of a possibly nonstationary process; in contrast, the bispectrum describes the *third*-moments of a stationary process and has an implicit third frequency $f_3 = f_1 + f_2$. Equation (3.21) highlights the underlying feature of a nonstationary process by describing the *correlation* between the spectral elements $X(f_1)$ and $X(f_2)$ of the process at two different frequencies f_1 and f_2, respectively.

If the process is stationary, then, by definition, the covariance $\Gamma_L(t_1, t_2)$ depends only on the time difference $t_1 - t_2$, and the Loève spectrum becomes $\delta(f_1 - f_2)S(f_1)$, where $\delta(t)$ is the Dirac delta function in the frequency domain and $S(f)$ is the ordinary power spectrum. Similarly, for a white nonstationary process, the covariance function becomes $\delta(t_1 - t_2)P(t_1)$, where $\delta(t)$ is the Dirac delta function in the time domain and $P(t)$ is the expected (average) power of the process at time t. Thus, as both the spectrum and covariance functions include delta-function discontinuities in simple cases, neither should be expected to be "smooth"; and *continuity properties* of the process therefore depend on direction in the (f_1,f_2) or (t_1, t_2) plane. The continuity problems are more easily dealt with by rotating both the time and frequency coordinates of the covariance function (3.20) and Loève spectrum (3.21), respectively, by 45°. In the time domain, we may now define the new coordinates to be a "center" t_0 and a delay τ, as shown by

$$\begin{aligned} t_1 + t_2 &= 2t_0 \\ t_1 - t_2 &= \tau. \end{aligned} \tag{3.22}$$

Correspondingly, we may write

$$\begin{aligned} t_1 &= t_0 + \tau/2 \\ t_2 &= t_0 - \tau/2. \end{aligned} \tag{3.23}$$

Thus, denoting the new covariance function in the *rotated coordinates* by $\Gamma(\tau, \tau_0)$, we may go on to write

$$\Gamma_L(t_1, t_2) = \Gamma(\tau, t_0). \tag{3.24}$$

Similarly, we may define new frequency coordinates, f and g, by writing

$$\begin{aligned} f_1 + f_2 &= 2f \\ f_1 - f_2 &= g. \end{aligned} \tag{3.25}$$

Correspondingly, we have

$$\begin{aligned} f_1 &= f + g/2 \\ f_2 &= f - g/2. \end{aligned} \tag{3.26}$$

The *rotated two-frequency spectrum* is thus defined by

$$\gamma(f, g) = \gamma_L(f + g/2, f - g/2). \tag{3.27}$$

Substituting the definitions of (3.26) into (3.20) shows that the term $(t_1f_1 - t_2f_2)$ in the exponent of the Fourier transform becomes $t_0g + \tau f$; hence, we have

$$\Gamma(\tau, t_0) = \int_{-\infty}^{\infty} \left\{ \int_{-\infty}^{\infty} \exp(j2\pi(\tau f + t_0 g)) \right\} \gamma(f, g) \mathrm{d}f \mathrm{d}g. \tag{3.28}$$

In view of the principle of duality that embodies the inverse relationship between time and frequency (an inherent characteristic of Fourier transformation), the frequency f is associated with the time difference τ; accordingly, f corresponds to the ordinary frequency of stationary processes; we may therefore refer to f as the "stationary" frequency. Similarly, the frequency g is associated with the average time t_0; therefore, it describes the behavior of the spectrum over long time spans; hence, we refer to g as the "nonstationary" frequency.

Consider next the continuity of the *generalized spectral density*, γ, reformulated as a function of f and g. On the line $g = 0$, the generalized spectral density γ is just the ordinary spectrum with the usual continuity (or lack thereof) conditions normally applying to stationary spectra. As a function of g, however, we expect to find a δ-function discontinuity at $g = 0$ if, for no other reason, that almost all data contain some stationary additive noise. Consequently, smoothers in the (f, g) plane or, equivalently, the (f_1,f_2) plane should not be isotropic but require much higher resolution along the nonstationary frequency coordinate g than along the stationary frequency axis f.

A slightly less arbitrary way of handling the g coordinate is to apply the inverse Fourier transform to $\gamma(f, g)$ with respect to the nonstationary frequency, g, obtaining [112]

$$D(t_0,f) = \int_{-\infty}^{\infty} \exp(j2\pi t_0 g)\gamma(f, g)\mathrm{d}g \tag{3.29}$$

as the *dynamic spectrum* of the process; the $D(t_0,f)$ in (3.29) is not to be confused with the time-frequency distribution in (3.17) and (3.18). The motivation behind (3.29) is to transform very rapid variations expected around $g = 0$ into a slowly varying function of t_0 while, at the same time, leaving the usual dependence on f intact. From Fourier-transform theory, we know that the Dirac delta function in the frequency domain is transformed into a constant in the time domain. It follows therefore that, in a stationary process, $D(t_0,f)$ does not depend on t_0 and assumes the simple form $S(f)$. Thus, we may invoke the Fourier transform to redefine the dynamic spectrum as

$$D(t_0,f) = \int_{-\infty}^{\infty} \exp(-j2\pi\tau f)\mathbf{E}\left\{x\left(t_0 + \frac{\tau}{2}\right)x^*\left(t_0 - \frac{\tau}{2}\right)\right\} \mathrm{d}\tau, \tag{3.30}$$

where the expectation, or ensemble averaging, is performed on the $x(t)$ for prescribed values of time t_0 and frequency f.

From an engineering perspective, we usually like to have estimates of second-order statistics of the underlying physics responsible for the generation of a nonstationary process. Moreover, it would be desirable to compute the estimates using the MTM. With this twofold objective in mind, let $X_k(f_1)$ and $X_k(f_2)$ denote the multitaper

Fourier transforms of the sample function $x(t)$; these two estimates are based on the kth Slepian taper and are defined at two different frequencies, f_1 and f_2, in accordance with (3.1). To evaluate the *spectral correlation* of the process at f_1 and f_2, the traditional formulation is to consider the product $X_k(f_1)X_k^*(f_2)$ where, as before, the asterisk in $X_k^*(f_2)$ denotes complex conjugation. Unfortunately, we often find that such a formulation is *insufficient* in capturing the underlying second-order statistics of the process, particularly so in the case of several communication signals that are of interest in cognitive-radio applications. For most complex-valued signals, the expectation $\mathbf{E}[x(t_1)x(t_2)]$, and therefore $\mathbf{E}[X(f_1)X(f_2)]$, is zero. For communication signals, however, this expectation is often not zero; examples of signals for which this statement holds include the ATSC-DTV signal, binary phase-shift keying (BPSK), minimum-shift keying (MSK), offset quadrature phase-shift keying (OQPSK), orthogonal frequency-division multiplexing (OFDM), and Gaussian minimum-shift keying (GMSK) used in Global System for Mobile communication (GSM) wireless communications. To complete the second-order statistical characterization of the process, we need to consider products of the form $X_k(f_1)X_k(f_2)$, which do *not* involve the use of complex conjugation. However, in the literature on stochastic processes, statistical parameters involving products such as $X_k(f_1)X_k(f_2)$ are frequently not named and therefore hardly used; and when they are used, not only different terminologies are adopted but also some of the terminologies are misleading.

In a terminological context, there is confusion in how second-order moments of complex-valued stochastic processes are defined in the literature:

- Thomson [101] and Picinbono [113] use the terms forward and reversed to distinguish, for example, the second-order moments $\mathbf{E}[X_k(f_1)X_k^*(f_2)]$ and $\mathbf{E}[X_k(f_1)X_k(f_2)]$, respectively.
- In [114], Moores applies spectral analysis to physical-oceanographic data, in the context of which two kinds of cross-correlation functions for a pair of complex-valued time series, $x_j(t)$ and $x_k(t)$, are introduced:
  1. The inner cross-correlation function is defined by the expectation $\mathbf{E}\{x_j^*(t)x_k(t+\tau)\}$ for some τ, where the asterisk denotes complex conjugation; this second-order moment is so called because it resembles an inner product.
  2. The outer cross-correlation function is defined by the expectation $\mathbf{E}\{x_j(t)x_k(t+\tau)\}$, where there is no complex conjugation; this alternative second-order moment is so called because it resembles an outer product.
- In the cyclostationarity literature on communication signals, the terms spectral correlation and conjugate spectral correlation are used to refer to the expectation $\mathbf{E}[X_k(f_1)X_k^*(f_2)]$ and $\mathbf{E}[X_k(f_1)X_k(f_2)]$, respectively. This terminology is misleading: If $\mathbf{E}[X_k(f_1)X_k^*(f_2)]$ stands for spectral correlation, then the expression for conjugate spectral correlation would be $\mathbf{E}[X_k^*(f_1)X_k(f_2)]$, which is not the intention.

Here, Moores' terminology is followed.

To put matters right, here, we follow the terminology first described in a 1973 paper by Moores [114], Middleton [115], and use the subscripts *inner* and *outer* to distinguish between spectral correlations based on products involving such terms as $X_k(f_1)X^*(f_2)$ and $X_k(f_1)X(f_2)$, respectively. Hereafter, estimates of spectral correlations so defined are referred to as estimates of the *first* and *second kinds*, respectively, and likewise for related matters.

With the matter of terminology settled, taking the complex demodulates of a nonstationary process at two different frequencies, f_1 and f_2, and invoking the inherent orthogonality property of Slepian sequences, we may now formally define the estimate of the Loève spectrum of the first kind as

$$\hat{\gamma}_{L,\text{inner}}(f_1,f_2) = \frac{1}{K}\sum_{k=0}^{K-1} X_k(f_1)X_k^*(f_2), \tag{3.31}$$

where, as before, K is the total number of Slepian tapers. The estimate of the Loève spectrum of the second kind is correspondingly defined as

$$\hat{\gamma}_{L,\text{outer}}(f_1,f_2) = \frac{1}{K}\sum_{k=0}^{K-1} X_k(f_1)X_k(f_2). \tag{3.32}$$

Thus, given a stochastic process with the complex demodulates $X_k(f_1)$ and $X_k(f_2)$, the *Loève spectral coherence of the first* and *second kinds* are, respectively, defined as

$$C_{\text{inner}}(f_1,f_2) = \frac{\hat{\gamma}_{L,\text{inner}}(f_1,f_2)}{(\hat{S}(f_1)\hat{S}(f_2))^{1/2}} \tag{3.33}$$

and

$$C_{\text{outer}}(f_1,f_2) = \frac{\hat{\gamma}_{L,\text{outer}}(f_1,f_2)}{(\hat{S}(f_1)\hat{S}(f_2))^{1/2}} \tag{3.34}$$

with the eigenvalue $\lambda_k(f)$ being real valued for all k and f, the multitaper spectral estimate $\hat{S}(f)$ in (3.3) is real valued, so it should be. In general, the Loève spectral coherences $C_{\text{inner}}(f_1,f_2)$ and $C_{\text{outer}}(f_1,f_2)$ are both complex valued, which means that each one of them will have its own magnitude and associated phase. The magnitudes of both spectral coherences are invariant under coordinate rotation, which is equivalent to multiplying $x(t)$ by $\exp(j\theta)$, where the constant θ is the angle of rotation. On the other hand, the phases of the inner and outer spectral coherences are altered by different amounts. In practice, we find that a quantity called the *two-frequency magnitude-squared coherence* (TF-MSC) is more useful than the spectral coherence itself. With the two spectral coherences of (3.33) and (3.34) at hand, we have two TF-MSCs to consider, namely, $|C_{\text{inner}}(f_1,f_2)|^2$ and $|C_{\text{outer}}(f_1,f_2)|^2$, respectively.

From the defining equation (3.30), we immediately recognize that

$$W(t_0,f) = \int_{-\infty}^{\infty} \exp(-j2\pi\tau f)x\left(t_0+\frac{\tau}{2}\right)x^*\left(t_0-\frac{\tau}{2}\right)\mathrm{d}\tau \tag{3.35}$$

is the formula for the *Wigner–Ville distribution* of the original sample function $x(t)$. In other words, we see that *the rotated Loève spectrum is the expected value of the Wigner–Ville distribution* [109, 112]. Stated in another way, the Wigner–Ville distribution is the *instantaneous estimate of the dynamic spectrum* of the nonstationary signal $x(t)$, and therefore simpler to compute than $D(t_0,f)$ in the classification of signals.

A cautionary note on the use of (3.35): The naive implementation of the Wigner–Ville distribution, as defined in this equation using a finite sample size, may result in bias and sampling properties that are worse than the periodogram. An improved version of the Wigner–Ville distribution was proposed in [116].

The dynamic spectrum also embodies another special case, namely, the *cyclic power spectrum* of a sample function $x(t)$ that is known to be *periodic*. Let T_0 denote the period of $x(t)$. Then, replacing the time t_0 in (3.30) with $T_0 + t$, we may express the time-varying power spectrum of $x(t)$ as

$$\begin{aligned} S_x(t,f) &= \int_{-\infty}^{\infty} \exp(-j2\pi\tau f)\mathbf{E}\left\{x\left(t+T_0+\frac{\tau}{2}\right)x^*\left(t+T_0-\frac{\tau}{2}\right)\right\}\mathrm{d}\tau \\ &= \int_{-\infty}^{\infty} \exp(-j2\pi\tau f)R_x\left(t+T_0+\frac{\tau}{2}, t+T_0-\frac{\tau}{2}\right)\mathrm{d}\tau, \end{aligned} \tag{3.36}$$

where

$$R_x\left(t+T_0+\frac{\tau}{2}, t+T_0-\frac{\tau}{2}\right) = \mathbf{E}\left[x\left(t+T_0+\frac{\tau}{2}\right)x^*\left(t+T_0-\frac{\tau}{2}\right)\right] \tag{3.37}$$

is the *time-varying autocorrelation function* of the signal $x(t)$. The stochastic process, represented by $x(t)$, is said to be *cyclostationary in the second-order sense* if this autocorrelation sequence is itself periodic with period T_0, as shown by

$$R_x\left(t+T_0+\frac{\tau}{2}, t+T_0-\frac{\tau}{2}\right) = R_x\left(t+\frac{\tau}{2}, t-\frac{\tau}{2}\right). \tag{3.38}$$

Under this condition, (3.36) reduces to

$$S_x(t,f) = \int_{-\infty}^{\infty} \exp(-j2\pi\tau f)R_x\left(t+\frac{\tau}{2}, t-\frac{\tau}{2}\right)\mathrm{d}\tau, \tag{3.39}$$

which, as expected, is independent of the period T_0. Equation (3.39) is recognized as the cyclostationary extension of the *Wiener–Khinchin relation* for stochastic processes.

To be more complete, for a stochastic process to be cyclostationary in the second-order sense, its mean must also be periodic with the same period, T_0. When the mean of the stochastic process under study is zero for all time t, this condition is immediately satisfied.

Before proceeding to discuss cyclostationarity characterization of nonstationary processes in the following section, we find it instructive to have a diagrammatic

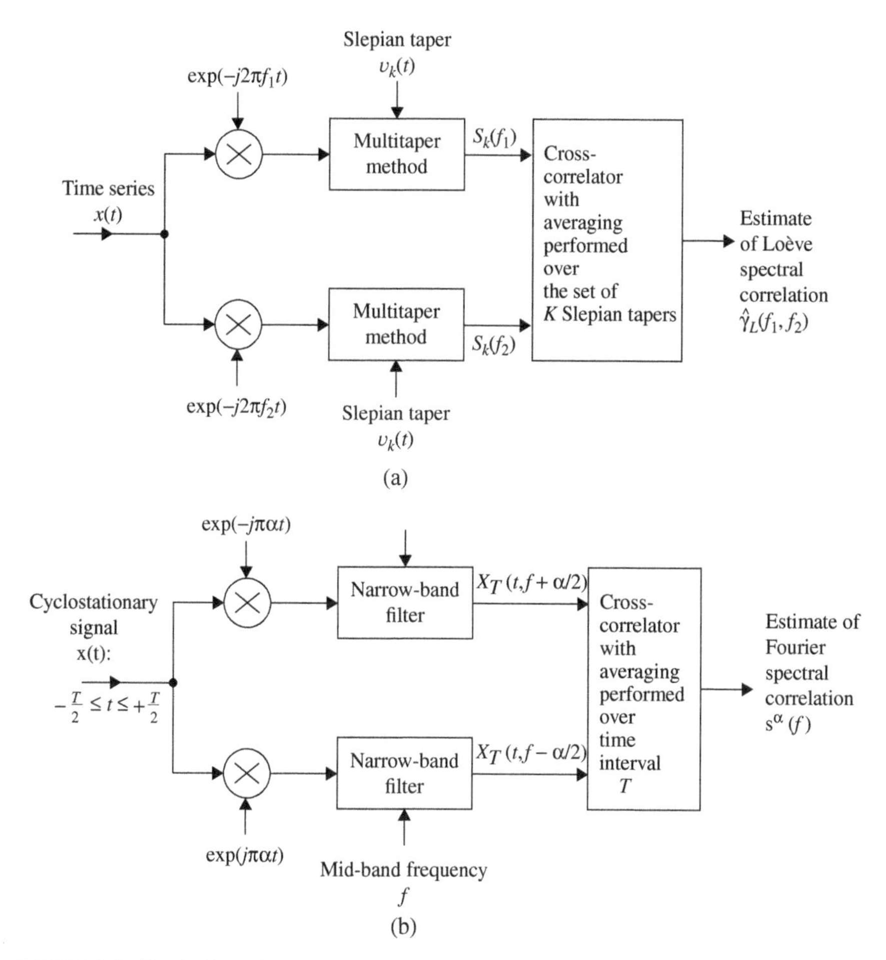

FIGURE 3.2 Illustrating the one-to-one correspondences between the Loève and Fourier theories for cyclostationarity. Basic instrument for estimating (a) the Loève spectral correlations of a time series $x(t)$ and (b) the Fourier spectral correlations of cyclostationary signal $x(t)$. Source: Haykin (2009) [48]. Reproduced with the permission of IEEE.

instrumentation for computing the Loève spectral correlations using the MTM. To do this, we look to the defining equations (3.1), (3.31), and (3.32), where t in (3.1) denotes discrete time and f in all three equations denotes continuous frequency. Let $x(t)$ denote a time series of length N. Then, the inspection of (3.1), (3.31), and (3.32) leads to the *basic instrument* diagrammed in Figure 3.2(a). In particular, in accordance with (3.1), the identical functional blocks labeled "multitaper method" in the upper and lower paths of the figure produce the Fourier transforms $X_k(f_1)$ and $X_k(f_2)$, respectively. The designation "basic" is intended to signify that the instrument applies to both kinds of the Loève spectral correlation, depending on how the cross-correlation of the Fourier transforms $X_T(f_1)$ and $X_T(f_2)$ is computed over the set of K Slepian tapers. To be specific, we say that as the overall output:

- the instrument computes the estimate $\hat{\gamma}_{L,\mathrm{inner}}(f_1,f_2)$ of (3.31) if the cross-correlation is of the first kind; and
- it computes $\hat{\gamma}_{L,\mathrm{outer}}(f_1,f_2)$ of (3.32) if the cross-correlation is of the second kind.

Figure 3.2(b) applies to spectral correlations rooted in the Fourier framework, considered in the following section.

3.1.5 Cyclostationarity: Fourier Perspective

A stochastic process represented by the sample function $x(t)$ is said to be *cyclostationary in the second-order sense* if its time-varying autocorrelation function $R_x(t+\tau/2, t-\tau/2)$ satisfies the periodicity condition of (3.38). Moreover, if the mean of the process is nonzero, it would also have to be time varying with the same period T_0. For the present discussion, the mean is assumed to be zero for all time t, and attention is therefore focused on second-order statistics.

A cyclostationary process may also be described in terms of its power spectrum, which assumes a periodic form of its own. With interest focused on spectral coherence, we now go on to define the inner and outer forms of spectral coherence of a cyclostationary process using Fourier theory.

Let $x(t)$ denote a sample function of a complex-valued cyclostationary process with period T_0. Using the Fourier series, we may characterize the process by its *cyclic power spectrum of the first kind*, as shown by the Fourier expansion:

$$S_{\mathrm{inner}}(t,f) = \sum_{\alpha} s^{\alpha}_{\mathrm{inner}}(f)\exp(j2\pi\alpha t), \tag{3.40}$$

where the new parameter α, in theory, scans an infinite set of frequencies, namely, n/T_0 where $n = 0, 1, 2, \ldots$. The power spectrum of (3.40) is cyclic in that it satisfies the condition of periodicity:

$$S_{\mathrm{inner}}(t+T_0,f) = S_{\mathrm{inner}}(t,f).$$

The Fourier coefficients in (3.40), namely, $s^{\alpha}_{\mathrm{inner}}(f)$ for varying α, are defined by

$$s^{\alpha}_{\mathrm{inner}}(f) = \lim_{T\to\infty}\lim_{\Delta t\to 0}\frac{1}{\Delta t}\int_{-\Delta t/2}^{\Delta t/2}\frac{1}{T}X_T(t,f+\alpha/2)X_T^{*}(t,f-\alpha/2)\mathrm{d}t. \tag{3.41}$$

The infinitesimally small Δt is included in (3.41) to realize the continuous-time nature of the cyclostationary signal $x(t)$ in the limit as Δt approaches zero. The *time-varying Fourier transform* of $x(t)$, denoted by $X_T(t,f)$, is defined by

$$X_T(t,f) = \int_{t-T/2}^{t+T/2} x(\xi)\exp(-j2\pi f\xi)\mathrm{d}\xi. \tag{3.42}$$

Most importantly, $s^{\alpha}_{\text{inner}}(f)$ is the time-average of the inner product $X_T(f+\alpha/2)X_T^*(f-\alpha/2)$; it follows therefore that $s^{\alpha}_{\text{inner}}(f)$ is the inner spectral correlation of the cyclostationary signal $x(t)$ for the two frequencies $f_1 = f+\alpha/2$ and $f_2 = f-\alpha/2$.

Equations (3.40) and (3.41) provide a *partial description* of the second-order statistics of a complex-valued cyclostationary process. To complete the statistical description, we need to introduce the *cyclic power spectrum of the second kind*, as shown by

$$S_{\text{outer}}(t,f) = \sum_{\alpha} s^{\alpha}_{\text{outer}}(f)\exp(j2\pi\alpha t), \tag{3.43}$$

where $s^{\alpha}_{\text{outer}}(f)$ is time average of the outer product $X_T(t,f+\alpha/2)X_T(t,f-\alpha/2)$, which does not involve the use of complex conjugation.

With the formulas of (3.41) and (3.43) at hand, we may now define the two *Fourier spectral coherences* of a cyclostationary process as follows:

1. *Fourier spectral coherence of the first kind*:

$$C^{\alpha}_{\text{inner}}(f) = \frac{s^{\alpha}_{\text{inner}}(f)}{(s^0(f+\alpha/2)s^0(f-\alpha/2))^{1/2}}. \tag{3.44}$$

2. *Fourier spectral coherence of the second kind*:

$$C^{\alpha}_{\text{outer}}(f) = \frac{s^{\alpha}_{\text{outer}}(f)}{(s^0(f+\alpha/2)s^0(f-\alpha/2))^{1/2}}. \tag{3.45}$$

Both spectral coherences have the same denominator, where the Fourier coefficient $s^0(f)$ corresponds to $\alpha = 0$; putting $\alpha = 0$ in the expressions for $s^{\alpha}_{\text{inner}}(f)$ and $s^{\alpha}_{\text{outer}}(f)$ yields the common formula:

$$s^0(f) = \lim_{T\to\infty}\lim_{\Delta t\to 0}\frac{1}{\Delta t}\cdot\int_{-\Delta t/2}^{\Delta t/2}\frac{1}{T}|X_T(t,f)|^2\,dt \tag{3.46}$$

As with the Loève spectral coherences, the Fourier spectral coherences are both complex valued in general, with each one of them having a magnitude and associated phase of its own.

The use of the Fourier spectral coherence of the first and second kinds in (3.44) and (3.45) can require excessive memory and therefore be computationally demanding in practice. To simplify matters, the *cycle frequency-domain profile* (CFDP) versions of spectral coherence are often used:

$$\text{CFDP}_{\text{inner}}(\alpha) = \max_{f}|C^{\alpha}_{\text{inner}}(f)| \tag{3.47}$$

and similarly for the outer spectral coherence $C^{\alpha}_{\text{outer}}(f)$.

The block diagram of Figure 3.2(b) depicts the instrument [117] for computing the inner and outer kinds of the Fourier spectral correlations at frequencies $f_1 = f+\alpha/2$

and $f_2 = f - \alpha/2$ in accordance with (3.41) for $s^{\alpha}_{\text{inner}}(f)$ and its counterpart for $s^{\alpha}_{\text{outer}}(f)$. A cyclostationary signal $x(t)$ is applied in parallel to two paths, both of which use identical narrowband filters. Both filters have the midband frequency f and bandwidth Δf, where the Δf is small compared with f but large enough compared with the reciprocal of the time T that spans the total duration of the input signal $x(t)$. In any event, the Fourier transform of the input $x(t)$ is shifted due to the multiplying factors $\exp(\pm j\pi\alpha t)$, producing the following filter outputs: $X_T(f + \alpha/2)$ in the upper path and $X_T(f - \alpha/2)$ in the lower path. Depending on how these two filter outputs are processed by the spectral correlator, the overall output is $s^{\alpha}_{\text{inner}}(f)$ or $s^{\alpha}_{\text{outer}}(f)$.

Much of the communications literature on cyclostationarity and related topics such as spectral coherence differ from that on multitaper spectral analysis. Nevertheless, these two approaches to cyclostationarity characterization of an input signal are related. In particular, examining parts (a) and (b) of Figure 3.2, we see that the two basic instruments depicted therein are similar in signal-processing terms, in that they exhibit the following one-to-one correspondences:

1. The multiplying factors $\exp(-j2\pi f_1 t)$ and $\exp(-j2\pi f_2 t)$ in Figure 3.2(a) play similar frequency-shifting roles as the factors $\exp(j\pi\alpha t)$ and $\exp(-j\pi\alpha t)$ in Figure 3.2(b).
2. The MTM in Figure 3.2(a) for a prescribed Slepian taper and the narrowband filter in Figure 3.2(b) for prescribed mid-band frequency f and parameter α perform similar filtering operations.
3. Finally, the cross-correlator operates on the MTM outputs $X_k(f_1)$ and $X_k(f_2)$ in Figure 3.2(a) to produce estimates of the Loève spectral correlations, while the cross-correlator in Figure 3.2(b) operates on the filter outputs $X_T(t,f + \alpha/2)$ and $X_T(t,f - \alpha/2)$ to produce the Fourier spectral correlations with $f_1 = f + \alpha/2$ and $f_2 = f - \alpha/2$.

Naturally, the instruments depicted in Figure 3.2(a) and (b) differ from each other by the ways in which their individual components are implemented.

The theory of cyclostationarity presented in this section follows the framework originally formulated in Gardner [117]. This framework is rooted in the traditional Fourier-transform theory of stationary processes with an important modification: introduction of the parameter α (having the same dimension as frequency) in the statistical characterization of cyclostationary processes. Accordingly, the cyclic spectral features computed from this formula depend on how well the parameter α matches the underlying statistical periodicity of the original signal $x(t)$.

The other theory on cyclostationarity follows the framework originally formulated in Thomson [112]. This latter framework combines the following two approaches:

- The Loève transform for dealing with nonstationary processes
- The MTM for resolving the bias-variance dilemma through the use of Slepian sequences.

This two-pronged mathematically rigorous strategy for the TFA of nonstationary processes has a *built-in capability to adapt* to the underlying statistical periodicity of the signal under study. In other words, it is nonparametric and therefore robust.

The Fourier-based approach to cyclostationarity may also acquire an adaptive capability of its own. In many spectrum-sensing applications based on this approach, the Fourier spectral coherences of the first and second kinds, defined in (3.44) and (3.45), are computed over the entire spectral domain of interest, and the *actual* cycle frequencies (i.e., statistical periodicity of the signal) may thus be accurately estimated. Applicability of the Fourier-based cyclostationarity approach to spectrum sensing is thereby extended to a wide range of situations, ranging from completely blind (no prior knowledge of periodicity) to highly targeted ones (known periodicities with possible errors).

Summing up the basic similarities and differences between the Loève and Fourier theories of stochastic processes, we say the following:

- Both theories perform similar signal-processing operations on their inputs.
- The Fourier theory assumes that the stochastic process is cyclostationary, whereas the Loève theory applies to any nonstationary process regardless of whether it is cyclostationary or not.

In the use of cyclostationarity as a tool for signal detection and classification, there are several practical issues that may present challenges:

1. Communication systems have timing variations due to the imprecision of their clocks. In practice, this means that the signal is not truly cyclostationary, but it may be over some finite blocks of time. Long-duration averaging, however, tends to attenuate the spectral correlation feature when the time-varying clock randomizes the phase [118].
2. Channel effects such as Doppler shift and/or fading diminish the periodic nature of the signal phase transitions (e.g., modulation), and thus can also reduce the practical extent of data collection [119].
3. Not all signals can be classified with second-order cyclostationarity. For example, there is an ambiguity between various forms of pulse-amplitude modulation (PAM) in the cyclic spectrum [120]. This ambiguity can be overcome to some extent by exploiting higher-order cyclostationarity, such as *cyclic cumulants*, but estimation of higher-order moments is known to require substantially more data as well as complexity [121]. Note also that higher-order moments are very sensitive to outliers such as impulsive noise.
4. For a given modulated signal $x(t)$, computing a three-dimensional surface defined by $|C_x^{(\alpha)}(f)|$ for varying α and f is computationally intensive. However, in practice, it may not be necessary to compute the entire surface if assumptions about the operating band can be made, thereby reducing the region of computation. Furthermore, some computationally reduced algorithms have been developed to deal with this difficulty [122].

5. If there are several signals in the environment, then some pattern-recognition techniques are necessary to identify and sort out the myriad of features to determine the combination of signals that created those features.
6. The final issue pertains to highly filtered signals. As the pulse shaping becomes more aggressive to reduce bandwidth, cyclic features for many types of modulation tend to diminish, requiring even more data to be collected so as to reduce the variance of the cyclostationarity estimator to reveal weak features.

The issues described herein have been discussed in the literature in the context of the classical Fourier theory of cyclostationarity.

3.1.6 Rayleigh Fading Channels

Because mobile communications are one of the probable applications of cognitive radio, we must consider the spectrum-estimation problem in a *Rayleigh-fading channel*. To proceed, consider a data-transmission system for a Rayleigh-fading channel, where $X(t)$ denotes the transmitted signal, and $C(t)$ is a complex narrowband Gaussian process defining the instantaneous channel characteristics. The received signal is defined by

$$Y(t) = C(t) \cdot X(t) + N(t), \tag{3.48}$$

where $N(t)$ is stationary, complex, white Gaussian noise. Here we assume a *Jakes' model* [123] for the channel, that is,

$$R(\tau) = \mathbf{E}\{C(t)C^*(t+\tau)\} = J_0(2\pi f_d \tau), \tag{3.49}$$

where, as before, $\mathbf{E}$ is the statistical expectation operator, the mean-square transfer through the channel is normalized to unity, $J_0(\cdot)$ is the Bessel function of zero order, f_d is the Doppler frequency, and τ is the time delay. The Doppler frequency is $f_d = \left(\frac{v}{c}\right) f_c$ where v is the mobile velocity, f_c is the carrier frequency, and c is, as usual, the velocity of light. Denoting the wavelength by $\lambda = c/f_c$ and distance by $s = v\tau$, the argument of the Bessel function J_0 is just $2\pi\left(\frac{s}{\lambda}\right)$. Note that the covariance function $R(\tau)$ is simply a function of the distance moved during time τ measured in wavelengths. There are at least two problems with the Jakes' model. First, it assumes that the received signal is a result of signals reflected from a two-dimensional array of *random* scatterers, and a glance at many buildings, with their rows of uniformly spaced windows or wall panels, suggests that a diffraction grating might be a better model. Second, the Bessel function $J_0(\cdot)$ is a *band-limited* function and implies that the fading process is deterministic. It is, in fact, just the leading term of a series [124]; nevertheless, it is adequate to explain many of the commonly encountered fading problems in practice. The explanation for the ripples is relatively simple. Consider the model of (3.48) as part of a spectrum-estimation problem and write the eigencoefficients as follows, ignoring the additive noise in (3.48):

$$y_k(f) = \sum_{t=0}^{N-1} C(t)x(t)v_t^{(k)} \exp(-j2\pi ft). \tag{3.50}$$

Now, consider the data taper to be the product $C_k(t) = C(t)v_t^{(k)}$; so the estimated eigenspectra $\hat{S}_{C,k}(f)|y_k(f)|^2$ will be the true spectrum convolved with an equivalent spectral window. The Fourier transform of $J_0(2\pi f_d \tau)$ is defined on the infinite interval $(-\infty, \infty)$ by

$$\tilde{C}(f) = \begin{cases} \frac{1}{\pi}[f_d^2 - f^2]^{-\frac{1}{2}} & \text{for} \quad |f| < |f_d| \\ 0 & \text{otherwise.} \end{cases} \tag{3.51}$$

Equation (3.51) implies that the expected value of the spectrum will be that from nonfading conditions, $S_{nf}(f)$, convolved with $\tilde{C}(f)$. This convolution poses a problem because $\tilde{C}(f)$ goes to infinity at $f \pm f_d$, and the power in the sidelobes of the windows decays asymptotically as f^{-2}, which is a characteristic of QPSK signaling; hence, formal convergence is problematic. In particular, the sidelobes of the Slepian sequences are asymptotically of the form $c_k \text{sinc}(fT)$, where sinc(·) denotes the sinc function; so, when adjacent sidelobes are spaced by $2f_d$, that is, when $T \approx 1/(2f_d)$, we should expect the presence of large ripples in the spectrum. On shorter time intervals, the tendency for the Fourier transform to have a maximum at the Doppler frequency is still present. When the resolution of the spectral estimator becomes finer than the Doppler frequency, we begin to see ripples. We emphasize that these ripples are *not* sidelobes of the spectral windows *per se* because these windows can resolve the spectrum without adaptive weighting; the procedure, however, is sensitive.

3.1.7 Remarks on Nonparametric Spectrum Sensing

In finding and then exploiting spectrum holes in ATSC-DTV and commercial cellular bands, cognitive radio has the potential to solve the *radio-spectrum-underutilization* problem. However, when the following compelling practical issues are recognized:

- the notoriously unreliable nature of wireless channels due to complexity of the underlying physics of radio propagation [97],
- the uncertainties surrounding the availability of spectrum holes as they come and go [66], and
- the need for reliable communication whenever and wherever it is needed [30],

we begin to appreciate the research and development challenges involved in building and commercializing cognitive radio.

In signal-processing terms, spectrum sensing of the radio environment in an unsupervised manner is one of those challenges, on which the whole premise of cognitive radio rests. The MTM is a nonparametric (i.e., model independent) and, therefore, *robust spectral estimator* that offers the following attributes for solving the radio-spectrum-underutilization problem:

(i) *Resolution of the bias-variance dilemma*, which is achieved through the use of Slepian sequences with a unique and remarkable property:

> The Fourier transform of a Slepian sequence (window) has the maximal energy concentration inside a prescribed bandwidth under a finite sample-size constraint.

Simply put, there is no other window that satisfies this property. Of the various windows described in the literature, the Kaiser window [125] gives a good approximation to the zeroth order prolate spheroidal wave function (PSWF), that is, the Slepian taper $\{v_t^{(0)}\}_{t=1}^{N}$. This window is based on an analytic approximation due to Rice [126]. Unlike fixed-parameter windows such as the Hamming and Parzen windows (that correspond roughly to time-bandwidth products of $NW = 2$ and $NW = 4$, respectively), the Kaiser window has an adjustable parameter α, which is upper bounded by the time-bandwidth product $C_o = NW$. However, there does not appear to be a simple equivalent of the Slepian sequence with $k \geq 1$. Consequently, the MTM is an *accurate* spectral estimator, as evidenced by the experimental results on ATSC-DTV and generic land-mobile radio signals [87].

(ii) *Real-time computational feasibility*, which is realized through prior calculation and storage of the desired Slepian coefficients for a prescribed time-bandwidth product and use of a state-of-the-art FFT algorithm.

(iii) *Multidirectional listening capability*, which is achieved by incorporating the MTM into a *space-time processor* for estimating the unknown directions of arrival of interfering signals.

(iv) *Cyclostationarity*, the characterization of which is performed by expanding the MTM to embrace the *Loève transform* that accounts for nonstationarity of incoming RF stimuli in a mathematically rigorous manner; the property of cyclostationarity provides an effective approach for the reinforced detection of spectrum holes and the classification of communication signals.

The message to take from the combination of these four attributes is summed up as follows:

> The multitaper method (MTM) is a method of choice for nonparametric spectrum-sensor that is capable of detecting spectrum holes in the radio band, estimating the average power in each subband of the spectrum, providing a sense of direction for estimating the wavenumber spectrum of interferers, and outputting cyclostationarity characterization of the receiver input for signal detection and classification; just as importantly, computation of this overall spectrum-sensing capability is feasible in real time.

Cyclostationarity is an inherent property of digital modulated signals that exhibit *periodicity*. It manifests itself in TFA of such signals. The Loève theory of TFA paves the way for finding cyclostationarity in a signal through the combined use of two complementary spectral parameters, namely, the inner and outer spectral coherences. The spectral coherence is said to be of the inner kind when the spectral correlation in its numerator is defined as the expectation $\mathbf{E}[X_k(f_1)X_k^*(f_2)]$, in which the $X_k(f)$ is the multitaper Fourier transform of the input signal $x(t)$ at frequency f. It is said to be of the outer kind when its numerator is defined as the expectation $\mathbf{E}[X_k(f_1)X_k(f_2)]$, which does not involve complex conjugation. The Fourier theory of cyclostationarity exploits the periodic property of the autocorrelation function of a cyclostationary process, or its power spectrum. In its own way, the Fourier theory also leads to the formulation of spectral coherences of the inner and outer kinds, which are defined in a manner similar to their Loève counterparts. The Loève and Fourier theories of cyclostationarity are indeed related in signal-processing terms, as discussed previously. Perhaps the way to distinguish between them is to say that when a nonstationary processes is considered, the Loève theory applies to any such process, whereas the Fourier theory is restricted to a cyclostationary process.

The notoriously unreliable nature of wireless channels is attributed to the fading phenomenon in electromagnetic radio propagation. This important issue was addressed previously by focusing on *Rayleigh fading channels*, based on *Jakes' model*. Although this model is incomplete in theoretical terms, it is adequate to explain many of the fading problems commonly encountered in practice. An important point is that in a fading environment, the use of excessively long data blocks should be avoided.

3.1.8 Filter-Bank Implementation of the Multitaper Method

Farhang-Boroujeny [127] has shown that the underlying analytic theory of the MTM, leading to the derivation of the multitaper spectral estimator, can indeed be reformulated in the framework of filter banks. In effect, the orthogonal Slepian sequences (windows) can be viewed as an *orthogonal bank of eigenfilters*, hence the new terminology "filter-bank spectral estimator." Implementation of the filter bank involves the use of a prototype low-pass filter that realizes the zeroth band of the filter bank. The remaining bands in the filter bank are realized through the use of *polyphase modulation*. What is pleasing about the filter-bank approach of deriving the multitaper spectral estimator is not only its novelty but also the fact that the theory of filter banks is well known in signal-processing literature [128].

3.1.9 Cooperative Spectrum Sensing

As mentioned in the previous section, efficient spectrum sensing is critical for cognitive radios in the open spectrum regime. Due to path loss and shadowing, a cognitive radio may not be able to detect the presence of a primary user. Cooperative sensing has been proposed to address this *hidden terminal* issue. In [129], a cooperative game theoretic approach has been proposed for cooperative sensing in cognitive radio networks in order to autonomously adapt the network topology to environmental changes caused by mobility of the users. In the proposed approach, a group of users form a coalition. Each one of them monitors the environment independently and uses energy-detection method to find out if any primary-user transmitter is active in its neighborhood. Then all of them send the results of their detection operations to a central node in the coalition, which is called the *coalition head*, for decision fusion. The node with the highest detection resolution in the coalition is chosen as the coalition head. In this way, the node with the highest detection ability does not need to send its decision over the control channel and, therefore, the possibly most accurate detection result will not be distorted before the fusion process takes place. The coalition head uses a decision fusion rule based on logical OR operation to fuse all the detection results received from other members of the coalition. Then, it sends the outcome of the decision fusion process back to all other nodes in the coalition. Each cognitive radio tries to maximize its detection probability subject to a constraint on the probability of false alarm. Decisions for joining a coalition or leaving a coalition are made by cognitive radios based on this utility. Therefore, coalitions are formed, broken, and their sizes are changed based on the two operations of merging and splitting of users.

Independent disjoint coalitions may form in the network due to the cost of cooperation. However, the coordination and communication costs must be taken into account in such a model. Since the coalition head sends the decision fusion result to all members of the coalition, all of the nodes in a coalition have the same utility, which is equal to the utility of the coalition. Therefore, the spectrum-sensing game is a coalitional game with nontransferable utility. It is obvious that the coalition head has to consume part of its resources for decision fusion and sending the result to other nodes. In the proposed model, the coalition head receives the same payoff as the others and it is not clear why the coalition head should accept the extra burden without getting any extra credit.

The authors claimed that increasing the number of the members of a coalition decreases the probability of missing a primary user's presence and increases the probability of false alarm. Therefore, there is a trade-off between detection ability and false alarm probability. A *grand coalition* may not form and there is always a constraint on the maximum number of nodes in a coalition, which is dictated by the acceptable false alarm probability. The considered decision fusion logic is very simple and does not take account of accuracy of decisions made by different nodes. More sophisticated decision-fusion algorithms can decrease the probability of false alarm but at the expense of increasing the information exchange and computational complexity. The coordination and communication costs also play a key role in determining the size of coalitions, but they have not been considered in the model. The stability of coalitions highly depends on the relative locations of cognitive radios and primary-user transmitters [129].

3.2 DYNAMIC SPECTRUM MANAGEMENT

In cognitive radio, the DSM assumes the task of assigning available resource of spectrum holes to competing cognitive radio users in accordance with environmental constraints. Just as the base station lies at the very core of a traditional wireless network, so it is with the DSM in a cognitive radio network. The DSM algorithm described in this section is based on the method, which was proposed in [130]. Simply put, whenever and wherever a pair of cognitive radio users needs a wireless link to communicate, the DSM is equipped with a self-organized algorithm to search the radio environment and find a common spectrum hole that makes the communication link a practical reality, which is realized through the Hebbian learning process. In basic mathematical terms, the DSM problem may be viewed as a special case of the well-known *graph-theoretic coloring problem*, the optimal solution of which is unfortunately NP-hard [131]. To get around this theoretical difficulty, we appeal to *self-organized learning*, inspired by the pioneering work of the psychologist Hebb [132], that occupies an important role in the human brain. It was first described in [46], wherein the terminology *self-organized-dynamic spectrum management* (SO-DSM) was coined.

This novel way of thinking is based on the *Tsigankov–Koulakov model* [133] for map formation in the superior colliculus. To be specific, an *energy function*, E_c, is defined in terms of *spatial correlations*, and the cognitive radio network is modified in an iterative manner so as to minimize the spatial correlation process, which, in turn, leads to minimizing the corresponding energy function.

Typically, a brain-inspired *self-organizing map* uses a *Hebbian learning rule* that embodies self-amplification and correlative processing [132] with the aim of extracting communication patterns from the input signal. The network optimization is modified in accordance with the knowledge extracted from the radio environment. The relationship between the elements of an input signal vector $\mathbf{x}(n)$ and output signal $y(n)$ is defined by the finite summation:

$$y(n) = \sum_{i=1}^{m} w_i(n)x_i(n), \tag{3.52}$$

where $x_i(n)$ is the ith element of the $m \times 1$ input vector $\mathbf{x}(n)$ at time step n and $w_i(n)$ is the tap weight that couples $x_i(n)$ to $y(n)$. At each time step of the learning process, an appropriate adjustment $\triangle\, w_i(n)$ is applied to each synaptic weight $w_i(n)$, and the learning process stores in memory the knowledge gained from the input signal. The general form of the weighting adjustment is expressed as follows:

$$\triangle\, w_i(n) = F(y(n), x_i(n)), \tag{3.53}$$

where $F(\cdot,\cdot)$ is a nonlinear function. One of the simplest forms for this function is

$$F(y(n), x_j(n)) = \eta y(n)x_i(n), \tag{3.54}$$

where η is a *learning-rate parameter*, which is a positive constant. In applying the Hebbian learning rule, there is a tendency for the tap weights to grow without bounds, hence the need for the addition of some form of *normalization*. One mathematically convenient normalized modification of the learning rule is to write

$$w_i(n+1) = \frac{w_i(n) + \eta y(n)x_i(n)}{\sqrt{\sum_{j=1}^{m} (w_j(n) + \eta y(n)x_j(n))^2}}. \tag{3.55}$$

This normalization creates "competition" among the tap weights so as to stabilize the learning process.

3.2.1 The Tsigankov–Koulakov Model

In this model, there are two layers of neurons, each consisting of an $N \times N$ lattice of neurons (i.e., computational units). Initially, as shown in Figure 3.3(a), the neurons from retina (input) layer are randomly connected to neurons in the collicular (output) layer. The goal of the algorithm is to modify the neuronal connections so that, as shown in Figure 3.3(b), every neuron in the input layer is connected to its spatially corresponding neuron in the output layer [130]. The weights are modified according to Hebbian rule and the change in the weights $\Delta\, v_j(\vec{r})$ is proportional to correlation between activities in input cell a_j and output cell $A(\vec{r})$:

$$\Delta\, v_j(\vec{r}) = \eta A(\vec{r}) a_j = -\frac{\partial E_c}{\partial v_j(\vec{r})}, \tag{3.56}$$

which interprets the weight adjustment as a gradient descent in E_c. The map formation occurs through a stochastic minimization of E_c by repeating the following steps:

- Randomly select two unnecessarily adjacent cells i and j from the output layer.
- Switch the cells of the input layer connected to cells i and j with probability

$$P_{\text{switch}} = \frac{1}{1 + e^{\Delta E_c}}. \tag{3.57}$$

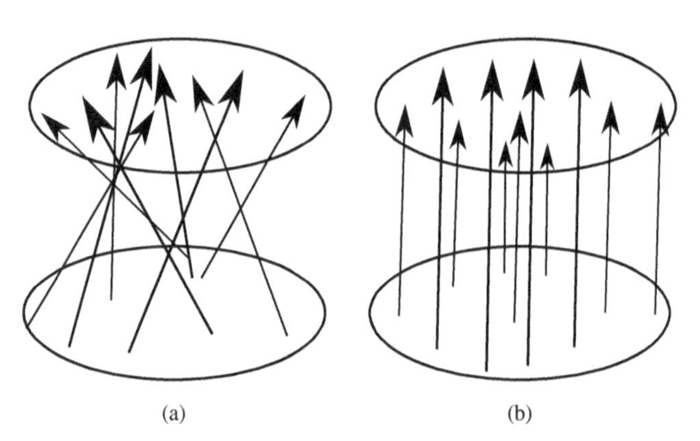

FIGURE 3.3 A self-organizing map: (a) initial state and (b) organized state. Source: Khozeimeh and Haykin (2012) [130]. Reproduced with the permission of IEEE.

Thus, this algorithm follows the gradient descent of E_c when cells are switched. According to equation (3.56), the energy function E_c decreases more if the correlation between input and output layers is higher, which happens if the spatial location of input cell in the input layer is close to the spatial location of output cell in the output layer [130].

3.2.2 Self-Organizing Dynamic Spectrum Management

Inspired by the human brain, the SO-DSM assigns the spectrum holes to secondary users based on the SOM technique [134]. This scheme is completely decentralized and each user communicates only with users in its neighborhood (i.e., other users in its communication range). Through local interactions, users exchange information to become coordinated, and eventually the network achieves a global organization in the following sense: Based on recent observations of the environment, each user builds a map of its surrounding environment and memorizes utilization patterns of the primary network and neighboring secondary users. When two secondary users need to establish a link, they choose the best channel based on their knowledge about those utilization patterns. Using this scheme, the cognitive radio network extracts both spatial and temporal patterns of primary users' activities and stores them in secondary users' memory.

We may therefore characterize a decentralized DSM scheme as one in which each secondary user extracts the wireless environment spectrum utilization patterns using Hebbian learning rule, storing the extracted knowledge in an array of weights. These weights act as short-term memory and keep recent spectrum utilization patterns of the legacy network and neighboring secondary users [130].

The prerequisite for forming an SOM is having correlation or redundancy in the input data, which is ordinarily the case. For SO-DSM, this redundancy is caused by the pattern of spectrum utilization of radio units in the environment. The wireless communications depends on humans' activities, which are not random and typically follow certain patterns. For example, in an office building, wireless activities increase at office hours during weekdays and significantly decrease during nights and weekends. Such a pattern for this environment is due to the fact that people are present in the environment mostly during office hours. Similarly, in a home environment, wireless activities increase during times when people are at home and not sleeping (i.e., weekday evenings and weekends), and decrease at nights and weekdays during the day when typically people leave home for school or work. Experimental measurements [135] have confirmed these kind of patterns and have shown that these patterns change at a relatively slow pace, in order of hours during the day [130].

We may therefore infer that wireless spectrum utilization has a pattern that depends on location, and the pattern may change slowly during the day [135]. Having a very fast radio-scene analyzer, we can consider the environment spectrum pattern to be *pseudostationary*, in which case the Hebbian learning rule can be applied to extract inherent patterns of the data. A good candidate for such radio-scene analyzer is the MTM, which was discussed previously. The presented results in [87] show that it is able to perform spectrum sensing relatively fast.

We may therefore assume that having such a fast radio-scene analyzer, the spectrum utilization pattern of environment is pseudostationary and the Hebbian learning rule can extract the patterns from the input data. We accomplish the other two requirements of SOM, namely self-amplification and competition, by using the Hebbian learning rule stated in equation (3.55). Finally, as SO-DSM is a decentralized cognitive radio network, the feedback channel between neighboring cognitive radio units is required to form the network.

The proposed SO-DSM technique works as follows [130]:

- Using the feedback channel on unlicensed band, the secondary users form an *ad hoc* network, become synchronized, and start sharing RSA information with their neighbors.
- Secondary users continuously monitor their surrounding environment and save the resulting information in a vector named *channel allocation priority list* (CAPL). This vector plays the role of short-term memory. Its size is equal to the total subbands, where for each subband k, secondary users store a weight w_k that represents the subband's quality in the recent past.
- After receiving each new set of RSA information, the weights are updated using equation (3.55).
- Once two secondary users need to establish a new link, subbands are allocated to links in the direction of gradient ascent of the utility function E_c as described in equation (3.56).

Simulation results presented in [130] followed the procedure just described and confirmed practical utility of the SO-DSM algorithm.

The above algorithm forms an SOM based on spectrum utilization patterns of primary users and cognitive radio units. However, it cannot meet one important requirement of cognitive radio units, which is minimizing probability of collision with primary users. This is due to the fact that using the above SOM model, the memory for activities of cognitive radio units and primary users is mixed and the weights show the *relative quality* of subbands. In other words, they are sorted based on their superiority with respect to each other and in effect therefore, these weights cannot be used to bound the probability of collision. To mitigate this problem, an extra stage was added to separate the memory for activities of primary users and cognitive radio units. Initially, all the weights are set to 0, so that the cognitive radio units do not use any spectrum holes before gaining enough knowledge about the environment. Subbands are considered *unavailable* as long as their corresponding weights are below 1. An unavailable subband will not be used for a cognitive radio link even if it is momentarily free (i.e., no primary user is using it). The memory weights for cognitive radio C_m, when they are in the unavailable state ($w_{i,m}(n) < 1$), are updated based on the following rule:

$$w_{i,m}(n+1) = \begin{cases} w_{i,m}(n) + \eta_{1,i,m}, & \text{if } b_i \text{ is free} \\ 0, & \text{if PU on } b_i, \end{cases} \tag{3.58}$$

where $1 \geq \eta_{1,i,m} > 0$ is the *forgetting factor* for subband b_i and cognitive radio unit C_m. When a weight reaches 1, its associated subband is considered *available* and will be used for cognitive radio communications. Using this rule, at each round, the weights for free subbands increase by $\eta_{1,i,m}$, and will eventually exceed 1 if no primary user uses them for at least $T_G = \lfloor \frac{1}{\eta_{1,i,m}} \rfloor$ consecutive time steps, where $\lfloor x \rfloor$ denotes the integer part of x. Once a primary user is detected on b_i, w_i is set back to zero and b_i will not be used by the cognitive radio unit, even if it becomes free, until it stays free for a sufficient amount of time to allow w_i to grow and reach 1.

After identifying the *available* subbands in the environment, each cognitive radio unit extracts the spectrum utilization pattern of neighboring cognitive radio units based on equation (3.55) and creates a temporal organization based on the obtained knowledge. Therefore, equation (3.55) is applied to the second stage of memory for updating the weights of available subbands ($w_{i,m}(n) \geq 1$):

$$w'_{i,m}(n+1) = \frac{w'_i(n) + \eta_{2,m} y(n) x_i(n)}{\sqrt{\sum_j (w'_j(n) + \eta_{2,m} y(n) x_j(n))^2}}, \tag{3.59}$$

where $1 > \eta_{2,m} > 0$ is the learning rate, $w'(n) = w(n) - 1$, and $x_{i,m}(n)$ is the quality signal of subband b_i received from RSA unit at time n, which is defined as

$$x_{i,m}(n) = \begin{cases} 1, \text{if } b_i \text{ is used by } C_m \\ \beta_1, \text{if } b_i \text{ is free} \\ \beta_2, \text{if } b_i \text{ is used by } C_{k,k \neq m} \end{cases}, \tag{3.60}$$

where $\beta_1 > \beta_2 > 0$ are quality levels for free subbands and subbands used by other cognitive radios. This quality signals represent the value or quality of the subband for each cognitive radio C_m based on its state. The state can be free, used by another cognitive radio, or used by C_m.

Using the Hebbian learning rule in equation (3.55), the utility function E_c exists [136]. Similar to the Tsigankov–Koulakov SOM model, we follow the gradient ascent direction for the utility function, ∇E_c, defined by equation (3.56), when assigning subbands to cognitive radio links. When a link between C_m and C_k is required, $b_j^* \in \mathcal{U}_a(k,m)$ is selected in a way that the utility function E_c is maximized, where

$$\begin{aligned} \mathcal{U}_a(k,m) &= \{b_j | b_j \text{ available for } C_k \text{ and } C_m\} \\ &= \{b_j \mid w_{j,m}, w_{j,k} \geq 1\}. \end{aligned} \tag{3.61}$$

The gradient of E_c is defined as

$$\nabla E_c(k,m,j) = \sum_{i=1}^{N_{\text{ch}}} \left(\frac{\partial E_c}{\partial w'_{i,m|j}} + \frac{\partial E_c}{\partial w'_{i,k|j}} \right),$$

where $w'_{i,m|j}$ is the next value of $w'_{i,m}$ if b_j gets assigned to the link between C_m and C_k. Using equation (3.56), we obtain

$$\nabla E_c(k,m,j) = \sum_{i=1}^{N_{\mathrm{ch}}} (\Delta w'_{i,m|j} + \Delta w'_{i,k|j}). \tag{3.62}$$

In order to calculate $w'_{i,k|j}$ for cognitive radio unit C_k, we use equation (3.55). By choosing b_j for the link, as equation (3.60) shows, only $x_{j,k}$ changes from β_1 to 1 and the rest of input signals remain the same as before, that is,

$$x_{i,k|j} = \begin{cases} x_{i,k}, & \text{if} \quad i \neq j \\ b' + x_{i,k}, & \text{if} \quad i = j \end{cases} \tag{3.63}$$

where $b' = 1 - \beta_1$. Substituting $x_{j,k}$ in equation (3.52), $y_{k|j}$ is calculated as

$$\begin{aligned} y_{k|j} &= \sum_{i=1}^{N_{\mathrm{ch}}} x_{i,k|j}\ w'_{i,k} = b'\ w'_{j,k} + \sum_{i=1}^{N_{\mathrm{ch}}} x_{i,k}\ w'_{i,k} \\ &= b'\ w'_{j,k} + y_k. \end{aligned} \tag{3.64}$$

Now, we can calculate $w'_{i,k|j}$ using equation (3.55):

$$w'_{i,k|j} = \frac{w'_{i,k} + \eta_{2,k}\ y_{k|j}\ x_{i,k|j}}{\sqrt{\sum_l (w'_{l,k} + \eta_{2,k}\ y_{k|j}\ x_{l,k|j})^2}}. \tag{3.65}$$

Then, we can expand the denominator of equation (3.65) using equations (3.64) and (3.63) as

$$\begin{aligned} \sum_l (w'_{l,k} + \eta_{2,k}\ y_{k|j}\ x_{l,k|j})^2 &= \sum_l (w'_{l,k} + \eta_{2,k}\ (y_k + w'_{j,k}\ b')\ x_{l,k|j})^2 \\ &= \sum_l (w'_{l,k} + \eta_{2,k}\ y_k\ x_{l,k|j})^2 + (\eta_{2,k}\ w'_{j,k} b'\ x_{l,k|j})^2 \\ &\quad + 2\eta_{2,k}\ w'_{j,k}\ b'\ x_{l,k|j}(w'_{l,k} + \eta_{2,k}\ y_k\ x_{l,k|j}) \\ &= \sum_l (w'_{l,k} + \eta_{2,k}\ y_k\ x_{l,k|j})^2 + 2\ \eta_{2,k}\ w'_{j,k}\ b'\ x_{l,k|j}\ w'_{l,k} \\ &\quad + 2\eta_{2,k}^2\ w'_{j,k}\ b'\ x_{l,k|j}^2\ y_k + \eta_{2,k}^2\ w'^2_{j,k}\ b'^2\ x_{l,k|j}^2, \end{aligned} \tag{3.66}$$

Having $\eta_{2,k} \ll 1$, $b' \ll 1$, and $w'_{j,k}, x_{l,k|j}, w'_{l,k} < 1$, we neglect the last three terms and obtain

$$\sum_l (w'_{l,k} + \eta_{2,k}\ y_{k|j}\ x_{l,k|j})^2 \approx \sum_l (w'_{l,k} + \eta_{2,k}\ y_k\ x_{l,k|j})^2 = \mathscr{D}^2. \tag{3.67}$$

Therefore, we can rewrite equation (3.65) as

$$w'_{i,k|j} = \frac{w'_{i,k} + \eta_{2,k}\,(y_k + w'_{j,k}\,b')\,x_{i,k|j}}{\mathcal{D}}. \tag{3.68}$$

Then, we will have

$$\begin{aligned}\Delta w'_{i,k|j} &= \frac{w'_{i,k} + \eta_{2,k}\,y_k\,x_{i,k|j} + \eta_{2,k}\,w'_{j,k}\,b'\,x_{i,k|j}}{\mathcal{D}} - w'_{i,k} \\ &= \mathcal{C}_{i,k|j} + \frac{\eta_{2,k}\,b'\,x_{i,k|j}}{\mathcal{D}} w'_{j,k},\end{aligned} \tag{3.69}$$

where

$$\mathcal{C}_{i,k|j} = \frac{w'_{i,k} + \eta_{2,k}\,y_k\,x_{i,k|j}}{\mathcal{D}} - w'_{i,k} \tag{3.70}$$

is the sum of terms in equation (3.69) that do not depend on $w'_{j,k}$. Using equation (3.63), we get

$$\begin{aligned}\mathcal{C}_{i,k|j} &= \begin{cases} \frac{w'_{i,k}+\eta_{2,k}\,y_k\,x_{i,k}}{\mathcal{D}} - w'_{i,k}, & \text{if} \quad i \neq j \\ \frac{w'_{i,k}+\eta_{2,k}\,y_k\,(x_{i,k}+b')}{\mathcal{D}} - w'_{i,k}, & \text{if} \quad i = j \end{cases} \\ &= \begin{cases} \mathcal{C}_{i,k}, & \text{if} \quad i \neq j \\ \mathcal{C}_{i,k} + \frac{\eta_{2,k}\,y_k\,b'}{\mathcal{D}}, & \text{if} \quad i = j \end{cases}\end{aligned}$$

Now we can calculate $\nabla E_c(k,m,j)$ using equations (3.62) and (3.69):

$$\begin{aligned}\nabla E_c(k,m,j) &= \sum_{i=1}^{N_{\text{ch}}} \left(\mathcal{C}_{i,k|j} + \frac{\eta_{2,k}\,b'\,x_{i,k|j}}{\mathcal{D}} w'_{j,k} + \mathcal{C}_{i,m|j} + \frac{\eta_{2,m}\,b'\,x_{i,m|j}}{\mathcal{D}} w'_{j,m} \right) \\ &= \sum_{i=1,\neq j}^{N_{\text{ch}}} \left(\mathcal{C}_{i,k} + \frac{\eta_{2,k}\,b'\,x_{i,k}}{\mathcal{D}} w'_{j,k} + \mathcal{C}_{i,m} + \frac{\eta_{2,m}\,b'\,x_{i,m}}{\mathcal{D}} w'_{j,m} \right) \\ &\quad + \mathcal{C}_{j,k} + \frac{\eta_{2,k}\,y_k\,b'}{\mathcal{D}} + \frac{\eta_{2,k}\,b'\,(x_{j,k}+b')}{\mathcal{D}} w'_{j,k} \\ &\quad + \mathcal{C}_{j,m} + \frac{\eta_{2,m}\,y_m\,b'}{\mathcal{D}} + \frac{\eta_{2,m}\,b'\,(x_{j,m}+b')}{\mathcal{D}} w'_{j,m} \\ &= \sum_{i=1}^{N_{\text{ch}}} \left(\mathcal{C}_{i,k} + \frac{\eta_{2,k}\,b'\,x_{i,k}}{\mathcal{D}} w'_{j,k} + \frac{\eta_{2,m}\,b'\,x_{i,m}}{\mathcal{D}} w'_{j,m} + \mathcal{C}_{i,m} \right) \\ &\quad + \frac{\eta_{2,k}\,y_k\,b'}{\mathcal{D}} + \frac{\eta_{2,k}\,b'^2}{\mathcal{D}} w'_{j,k} + \frac{\eta_{2,m}\,y_m\,b'}{\mathcal{D}} + \frac{\eta_{2,m}\,b'^2}{\mathcal{D}} w'_{j,m}\end{aligned}$$

$$= \sum_{i=1}^{N_{\text{ch}}} (\mathscr{C}_{i,k} + \mathscr{C}_{i,m}) + \frac{\eta_{2,k}\, y_k\, b'}{\mathscr{D}}$$

$$+ \frac{\eta_{2,m}\, y_m\, b'}{\mathscr{D}} + \frac{\eta_{2,k}\, b'}{\mathscr{D}} \left(b' + \sum_{i=1}^{N_{\text{ch}}} x_{i,k} \right) w'_{j,k}$$

$$+ \frac{\eta_{2,m}\, b'}{\mathscr{D}} \left(b' + \sum_{i=1}^{N_{\text{ch}}} x_{i,m} \right) w'_{j,m}. \tag{3.71}$$

Defining

$$\mathscr{S}_{x,k} = \sum_{i=1}^{N_{\text{ch}}} x_{i,k} \tag{3.72}$$

and

$$\mathscr{T}_k = \sum_{i=1}^{N_{\text{ch}}} \mathscr{C}_{i,k} + \frac{\eta_{2,k}\, y_k\, b'}{\mathscr{D}}, \tag{3.73}$$

we rewrite equation (3.71) as

$$\nabla E_c(k, m, j) = \mathscr{T}_k + \mathscr{T}_m + \eta_{2,k}\, b' \, \frac{\mathscr{S}_{x,k} + b'}{\mathscr{D}} w'_{j,k} + \eta_{2,m}\, b' \, \frac{\mathscr{S}_{x,m} + b'}{\mathscr{D}} w'_{j,m}. \tag{3.74}$$

Thus, the criteria to select b_j^* as the subband for cognitive radio units C_m and C_k is

$$b_j^* = \underset{j}{\operatorname{argmax}}(\nabla E_c(k, m, j)). \tag{3.75}$$

Equivalently, $\forall b_i \in \mathscr{U}_a(k, m)$, $i \neq j$, b_j^* must satisfy the following equation:

$$\frac{b'}{\mathscr{D}} [\eta_{2,k}\, (\mathscr{S}_{x,k} + b')\, (w'_{j,k} - w'_{i,k}) + \eta_{2,m}\, (\mathscr{S}_{x,m} + b')(w'_{j,m} - w'_{i,m})] \geq 0. \tag{3.76}$$

Assuming $\eta_{2,m} = \eta_{2,k} = \eta_2$, and eliminating the positive variables η_2, $\mathscr{D}$, and b', then, $\forall b_i \in \mathscr{U}_a(k, m)$, $i \neq j$ we obtain

$$(\mathscr{S}_{x,k} + b')(w'_{j,k} - w'_{i,k}) + (\mathscr{S}_{x,m} + b')(w'_{j,m} - w'_{i,m}) \geq 0. \tag{3.77}$$

Using equation (3.59), the weights approach the principal component of quality signals of subbands in the recent past [48]. At each step, if a subband is used by a cognitive radio unit, its weight increases more than other subbands' weights and gains higher priority in CAPL. Therefore, it is more likely that it will be used by that cognitive radio in the future. Similarly, weights of subbands that are being used by other cognitive radios would decrease; thus, they will go down in CAPL and therefore will be less likely to be used. However, as equation (3.77) shows, the link is selected based on the value of weights of both cognitive radio units that share the link. Therefore, it is possible that the selected subband is not on top of the CAPL of one of the cognitive radio units.

The learning rate η_1 defines the cognitive radio network behavior in its interaction with legacy networks. A larger η_1 results in faster forgetting primary user activities, thus, a more *aggressive* behavior, while a smaller η_1 results in remembering primary user activities for a longer time and therefore a more *conservative* behavior. By aggressive we mean cognitive radios wait for a shorter time after a primary user has stopped using a subband to use that subband. A larger η_1 would increase spectrum utilization of cognitive radio network while increasing the probability of collision with primary users. Cognitive radio units should adjust the forgetting factor to maintain the probability of collision at an acceptable level based on the legacy network traffic. Let $P_{s,i}(n)$ denote the probability that a primary user starts using b_i exactly at time n, then $P_{s,i}(n) = (1 - \mu_1)^{n-1}\mu_1$. If a neighboring cognitive radio unit C_j starts using b_i at time T_G for k time steps, then the probability of collision $P_{\text{col},i,j}$ will be

$$P_{\text{col},i,j} = D_{\text{RSA}} \times \sum_{T_G}^{T_G+k} P_{s,i}(n), \tag{3.78}$$

where D_{RSA} is the RSA time delay in discovering primary users. Note that $P_{s,i}(n)$ is a strictly decreasing function because

$$\frac{\mathrm{d}P_{s,i}(n)}{\mathrm{d}n} = \frac{\mu_1 \ln\,(1 - \mu_1)}{(1 - \mu_1)}, \tag{3.79}$$

which is always negative for $1 \geq \mu_1 > 0$. Therefore, it is possible to decrease the probability of collision by decreasing η_1 (i.e., increasing T_G). However, decreasing η_1 decreases the spectrum utilization of the cognitive radio network. In order to use the spectrum holes efficiently, cognitive radios need to have a good estimate of μ_1. They can obtain this information in two ways:

- *Using a stored look-up table of forgetting-factor profiles based on time and location, and finding the best match for their current condition* such as (university, daytime) or (home, night). As discussed earlier, wireless traffic is caused by human activities and has patterns that can be predicted to some extent, given the time and location.
- *Adaptively refining the learning rate based on recent observations of the environment*. The traffic patterns of primary users change slowly, in order of hours, during the day [137]. Using an algorithm such as exponential moving average (EMA) [138–140] or autoregressive models [141], which have been suggested for finding traffic parameters in the literature, cognitive radio units can estimate primary users' traffic parameters. We can make it even more refined through the use of hidden Markov model (HMM) [142], but the complexity may become unmanageable and it could very well be out of practical scope.

From a practical perspective, we suggest combining these two methods to achieve a more precise estimation in a shorter time interval. First using a look-up table, cognitive radio units can obtain a rough estimate, which can then be adaptively refined using the second method.

The second learning rate η_2 controls the sensitivity of the SOM algorithm to cognitive radio network changes. A higher η_2 would result in a more rapid response to a similar input than a lower η_2. If η_2 is too small, on the other hand, barely any learning or organization would happen and SO-DSM would basically assign subbands in a random manner. On the other hand, too large an η_2 would result in a very fast-learning and sensitive system, which would lose organization due to unnecessary responses to input data. Therefore, this parameter must be chosen according to the expected rate of changes in the environment.

There are two other parameters, which are important in shaping the weight dynamics in SO-DSM: β_1 and β_2. In general, we would like to have a large β_1 (i.e., close to 1) and a small β_2 (i.e., close to 0) to have more separation between signal levels for the subbands used by other cognitive radios and free subbands. Increased separation tends to increase the self-amplification of the Hebbian learning and helps forming the SOM. However, these parameters cannot get too close to the limits (i.e., 1 for β_1 and 0 for β_2).

This section presented the SO-DSM scheme for cognitive radio *ad hoc* networks. This scheme adds memory and learning, two essential elements of cognition, to radio units. In this scheme, cognitive radio units continuously monitor their environment, extract primary users' and neighboring cognitive radios' activity patterns, and store the obtained knowledge in their memory. When two cognitive radio units need to establish a link for communication, they use a common available subband that maximizes the link quality based on the stored knowledge. This scheme can significantly reduce the probability of collision and cognitive radio link interruption at the price of a moderate reduction in cognitive radio network spectrum utilization. The SO-DSM is a decentralized scheme with low complexity and minimal memory requirement. It is therefore suitable for mobile units with limited processing capability and battery life.

3.2.3 Dynamic Spectrum Management Based on Minority Games

Minority game is based on inductive learning, in which users try to predict the future of the network based on the history. The minority game introduces self-organized behavior in a decision-making process and models a situation, where being in the minority group is desirable. It is a branch of game theory that studies competition and self-imposed cooperation in a noncooperative game with limited resources [143]. Players in the original form of this game play according to a binary strategy set and do not directly interact or negotiate with each other. The original form of minority game, which is also known as the *El Farol Bar* problem, can be described as follows. A group of n players must decide independently at the same time whether or not to go to the El Farol Bar on Friday night. Hence, at each stage, a player has a binary strategy set: to go or not to go to the bar, where going to the bar would be enjoyable only if the bar is not too crowded. If the majority of players decide to go to the bar, then they will not enjoy as much as the other players, who decided to stay home. In other words, the players that are in the minority group will come out of the game as winners and in effect, therefore, players must do their best to be in the minority group [144].

In [145], minority game was proposed for channel assignment. Although only a couple of very simple cases were mentioned there, the idea is very interesting and it was worth to be expanded. If multichoice and multifrequency minority game is used for channel assignment, each channel will be shared among minimum number of users. This leads to a lower interference level in each channel and minimizes the total waste of network resources. A number of studies have already applied minority games to resource management in wireless communication systems [146–148].

Dynamic spectrum management can be performed based on the idea of minority games. In this framework, the cognitive radio transceiver pays attention to the history of activities in different subbands and tries to choose a set of subbands that had the lowest level of occupancy in the time interval of observation. In other words, memory may play an essential role here.

Let us assume that there are N active cognitive radio users in the network, who compete against each other for accessing M channels. At each round of the game, user i has to choose between one of the two actions regarding each channel, namely, select the channel or not, which are denoted by "1" and "0," respectively. Regarding channel k at time t, user i's action is denoted by $a_k^i(t)$. At each round, the minority choices win the game and the winning users are rewarded by consuming less power for transmitting over channels with lower levels of interference. In the beginning, each user may draw S strategies from a strategy pool. These strategies will be used as rough guidelines for decision-making during the game with no a priori knowledge about the best strategy. Such strategies may be viewed as tables with a history (i.e., signal) column and a prediction column. Each element of the history column is a string of ℓ bits representing the history of the winning actions in the last ℓ rounds of the game, which is called signal or information. The parameter ℓ represents the user's *memory size*, which is effective in the decision-making process. History evolves with time and is denoted by $\mu(t)$. In a game with the memory size of ℓ, the total number of possible signals is 2^ℓ; hence, the total number of possible strategies in the strategy pool will be 2^{2^ℓ}. Apparently, the total number of possible strategies will be huge even for relatively small memory size. Strategies may be represented by 2^ℓ-dimensional vectors that record predictions. If a strategy correctly predicts the winning action, it gains one reward point. All of the initially picked S strategies of a user predict the winning action for every channel at every round of the game. Then, reward points are given to strategies with correct predictions regardless of whether or not they were used by the user in the decision-making process for taking actions. The scores of all the strategies are accumulated, which are known as cumulated payoffs. Such payoffs start at zero in the beginning. At every round of the game, users make decisions using the strategy with the highest cumulated payoff at that time instant. If there is more than one strategy with the highest payoff, one of these strategies may be selected randomly. Users that come out of each round of the game as winners are also rewarded. These rewards are called the real points to be distinguished from the virtual points of the strategies.

In this framework, cognitive radios must be *adaptive* in the sense that they make decisions based on a strategy, which is chosen from their sets of S strategies. Such a selection occurs according to the relative preference of strategies, which in turn, is modified over the course of time taking account of the game outcomes at each round. Cognitive radios must also be *inductive* in the sense that they should make the best decisions they are aware of having a limited capacity for information gathering and processing. However, such decisions will not be the global best choice based on the strategy with the highest virtual score over the entire strategy space. Minority games provide a self-contained framework for dynamic spectrum management in the sense that users individually make decisions and take actions according to the history recorded in their memories. Subsequently, outcomes of such individual actions are summed up to make the history for the next round of the game. This updated history is then used by the cognitive radios to make predictions again [96].

3.2.4 Self-Organized Maps versus Minority Games

In [47], simulation results were provided to compare SO-DSM with minority game-based DSM (MG-DSM) regarding the following issues:

- Spectrum utilization
- Probability of collision for primary users and probability of link interruption for cognitive radio users
- Interference experienced by primary user receivers
- Mean and variance of subband assignment distribution.

According to the provided results, when the density of primary users is low, MG-DSM achieves better spectrum efficiency. However, by increasing the density of primary users, MG-DSM gradually loses its advantage and the performance gap between the two schemes decreases. In high densities, SO-DSM may even perform slightly better than MG-DSM. The SO-DSM outperforms MG-DSM by providing lower average probabilities for both primary user collisions and cognitive radio link interruptions. However, for primary user collisions, the performance gap decreases as the density of primary users increases. For low-density primary user networks, the SO-DSM outperforms the MG-DSM by causing lower levels of interference on primary user receivers but as the density of primary users increases, MG-DSM would slightly do better. Using SO-DSM, cognitive radio users in average will occupy a smaller number of subbands (i.e., lower average), which is an indication of a map formation. However, the higher variance of subband assignment distribution demonstrates that cognitive radio users can successfully build temporal organization.

It was concluded in [47] that adopting the SO-DSM scheme, cognitive radio transceivers will be more robust and provide more reliable communication links;

and by extracting the primary user communication patterns and forming a temporal subband organization, they will experience fewer interruptions from primary users and also impose less interference on them. The SO-DSM trades off a bit of spectrum efficiency for significantly less collisions and interruptions. However, for extreme scenarios with high-density primary and secondary networks, the SO-DSM achieves a slightly higher spectrum efficiency while still decreasing the probability of collisions and interruptions significantly.

3.3 TRANSMIT-POWER CONTROL

A cognitive radio must have the built-in ability to fill the spectrum holes rapidly and efficiently. In other words, cognitive radios have to be frequency-agile radios with flexible spectrum-shaping abilities. The OFDM scheme can provide the required flexibility and is therefore a good candidate for cognitive radio [30, 107, 149–151]. OFDM can be employed in a cognitive radio network by dividing the primary user's unused bandwidth into a number of subbands available for use by the cognitive radio systems. In order to achieve low mutual interference between primary and secondary users, an adaptive transmit filter can be used to prevent usage of a set of subcarriers, which are being used by the primary users. Moreover, the FFT block in the OFDM demodulator (Figure 3.4) can be used for spectral analysis [149]. The TPC algorithm described in this section is based on the method, which was proposed in [66].

OFDM is a multicarrier scheme in which a wideband signal is converted to a number of narrowband signals. Then closely spaced orthogonal subcarriers are used to transmit these narrowband data segments simultaneously. In effect, a frequency selective fading channel is divided into a number of narrowband flat-fading subchannels. OFDM has many advantages over single-carrier transmission [152–156]:

- It improves the efficiency of spectrum utilization by the simultaneous use of multiple orthogonal subcarriers, which are densely packed.
- The OFDM waveform is first built in the frequency domain and then it is transformed into the time domain, thereby providing flexible bandwidth allocation.
- *Interleaving* the information over different OFDM symbols provides robustness against loss of information caused by flat-fading and noise effects.
- Although the spectrum tails of subcarriers overlap with each other, at the center frequency of each subcarrier, all other subcarriers are zero. Theoretically, this prevents *intercarrier interference* (ICI). However, time and frequency synchronization is critical for ICI prevention as well as correct demodulation and is a major challenge in the physical layer design [157].
- Since a narrowband signal has a longer symbol duration than a wideband signal, OFDM takes care of *intersymbol interference* (ISI) caused by multipath delay of wireless channels. However, guard time intervals, which are longer than the

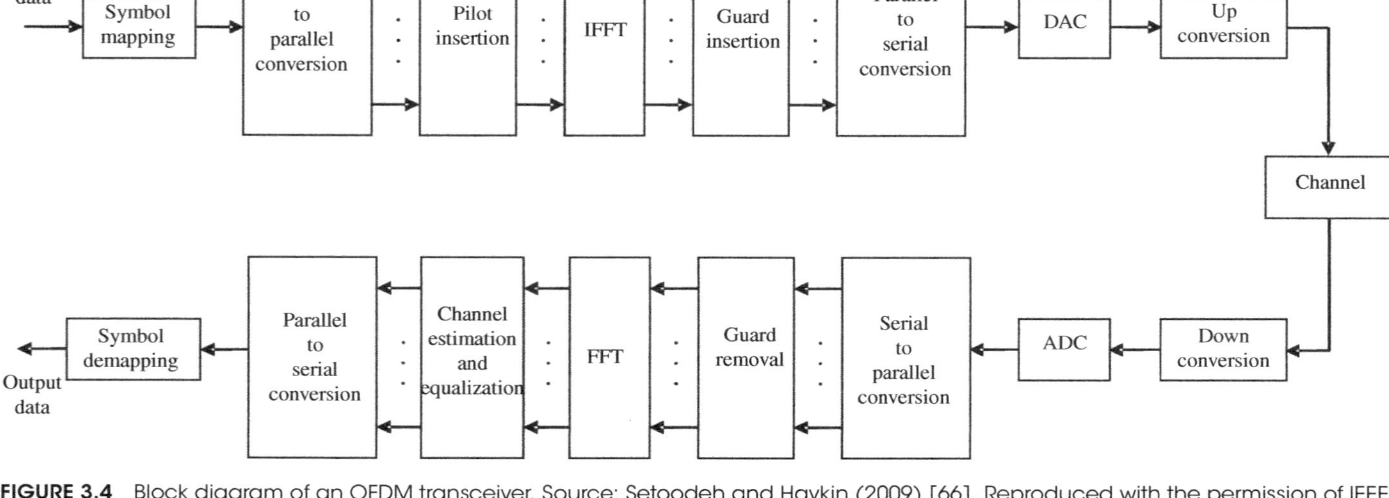

FIGURE 3.4 Block diagram of an OFDM transceiver. Source: Setoodeh and Haykin (2009) [66]. Reproduced with the permission of IEEE.

channel impulse response, are introduced between OFDM symbols to eliminate the ISI by giving enough time for each transmitted OFDM symbol to dissipate considerably [154].

- Due to the low ISI, less complex equalization is required at the receiver, which leads to a simpler receiver structure.

In summary, frequency diversity enables OFDM to provide higher data rates, more flexibility in controlling the waveform characteristics, and greater robustness against channel noise and fading compared to single-carrier transmission schemes.

Another appealing characteristic of OFDM is its low complexity of implementation. Low complexity is achieved, based on the assumption of subcarriers being a set of perfectly synchronized orthogonal tones that are generated by an inverse fast Fourier transform (IFFT) block in the transmitter and separated by an FFT block in the receiver. As mentioned previously, the FFT block in the OFDM demodulator can be used for spectral analysis as well.

Due to the advantages of OFDM, it was adopted as the modulation technique in the long-term evolution (LTE) fourth-generation (4G) cellular networks. Moreover, the chosen multiple-access strategy in LTE systems is the orthogonal frequency division multiple access (OFDMA) [158]. As an alternative approach, in a filter-bank multicarrier (FBMC) scheme, a bank of filters separates subcarriers. Hence, it would be compatible with an asynchronous operation. In other words, FBMC provides robustness to synchronization errors. However, unlike development of MIMO–OFDM, which is straightforward, the development of MIMO–FBMC is nontrivial [159]. Another weak point of OFDM is its probable high peak-to-average power ratio (PAPR). In order to limit the PAPR, single-carrier modulation (SCM) schemes are attracting attention again. To a large extent, the SCM techniques owe this regain of interest to frequency-domain equalization that can address the complexity issue, from which the SMC was suffering [160].

For the fifth generation (5G) cellular communications, in order to achieve higher spectral efficiency, it seems that none of the mentioned modulation schemes will be a definite winner. Hence, researchers may have to seek an adaptive modulation scheme with tunable waveform parameters [161]. In this regard, special attention must be paid to air interface virtualization and cloud radio access networks [161, 162].

In general, a cognitive radio can use a legacy user's subband if that subband can be divided into an integer number of cognitive radio's subcarriers. Using the OFDM-based modulation scheme, the bandwidth allocation can be considered as a subcarrier assignment problem [154]. The resource management problem may then consist of subcarrier assignment and power control. If cognitive radios and primary users use the same set of subcarriers, then they would have to be synchronized. In this case, the synchronization requirement would be a challenge for OFDM.

While the availability of channel bandwidth depends on the communication patterns of primary users, a cognitive radio has complete control over its own transmit power. In other words, among the two primary resources, power is the variable that can be manipulated by cognitive radio users. A subcarrier will not be assigned to a cognitive radio if its transmit power on that subcarrier is zero. Therefore, the resource-allocation problem can be reduced to the subcarrier assignment and transmit-power control over the assigned set of subcarriers, which can be considered as a distributed control problem. Scalable decentralized algorithms with reasonable computational complexity are naturally preferred.

In a cognitive radio network, the radio communication channel is shared between different transceivers and each user's action affects the performance of neighboring users while they compete for limited resources. At any instant of time, new users may join the network or old users may leave the network. Also, primary users may start or stop communication and, therefore, they may occupy or release some frequency bands in a stochastic manner. All of these occurrences can be considered as discrete events compared to the real-time evolution of each user's power vector, which can be considered as evolving in continuous time. It follows therefore that the cognitive-radio problem is a mixture of continuous dynamics and discrete events. In other words, a cognitive radio network is a hybrid dynamic system of the sort described in [163, 164].

The feedback channel will naturally introduce some delay in the control loop, and some of the users may use inaccurate or outdated interference measurements to update their transmit powers. Also, they may update their transmit powers with different frequencies. Therefore, in a real-life situation, the resource-allocation algorithm would have to be implemented in a distributed asynchronous manner [165–168].

In a competitive multiagent environment with limited resources such as a cognitive radio network, where the actions of all agents (users) are coupled via available resources, finding a global optimum for the resource-allocation problem can be computationally intractable and time consuming. Moreover, such optimization would require huge amounts of information exchange between different users that will consume precious resources. In a highly dynamic environment, where both users and resources can freely come and go, finding a reasonably good or "just right" solution (i.e., a suboptimal solution) that can be obtained fast enough is the only practical goal. Otherwise, spectrum holes may disappear before they can be utilized for communication. In such a situation, the concept of equilibrium is very important [169]. It is therefore not surprising that *game theory* has attracted the attention of many researchers in the field of communication networks. The references [170–174] are worth mentioning among the others for application of game theory in wireless communication systems and cognitive radio networks.

As mentioned previously, in game theory, the *Nash equilibrium* is a concept of fundamental importance. This equilibrium point is a solution such that none of the agents has an incentive to deviate from it unilaterally. In other words, in a Nash-equilibrium point, each user's chosen strategy is the "best response" to the other users' strategies [68, 74, 75]. Regarding the highly time-varying nature of communication networks in general and especially cognitive radio networks, a Nash-equilibrium solution is a reasonable candidate, even though it may not always be the best solution in terms of spectral efficiency [175].

The above discussion reveals that several key attributes such as distributed implementation, low complexity, and fast convergence to a reasonably good solution provide an intuitively satisfying framework for choosing and designing resource-allocation algorithms for cognitive radio. It is with this kind of framework in mind that in [30, 66, 107], the IWFA has been proposed as a good candidate for finding a Nash equilibrium solution for resource allocation in cognitive radio networks.

Regarding the coexistence of several transceivers, a set of constraints must be imposed on each user's transmit power in each subcarrier to maintain a limit on the interference produced. In [176], a fixed limit on each user's transmit power in each subcarrier is considered in order to guarantee that all users transmit at low powers and do not cause high interference. However, this approach may be too conservative from spectral efficiency point of view especially when a subband is not crowded. In [67, 177], global and flexible constraints were proposed instead of individual and rigid constraints. The peak average interference tolerable by the primary user's receiver is used to put a limit on cognitive radios' transmit powers. The measurements are performed at the primary user's receiver and the results are sent to secondary users' transmitters. This approach requires information exchange between primary users and secondary users and can be used in a market-model spectrum-sharing regime that involves pricing. In [66], the *permissible interference power level* limit (i.e., the *interference temperature* limit), which was proposed by FCC, was used as a local and flexible constraint. In the proposed approach, each user's receiver measures the interference power level on each subcarrier and sends the results to its corresponding transmitter through the feedback channel. The transmitter adjusts its transmit power vector in a way that it does not violate the permissible interference power level limit. As will be discussed in the following, the IWFA is a potentially good candidate for resource allocation in cognitive radio networks because of its low complexity, fast convergence, distributed nature, and convexity.

3.3.1 Waterfilling Interpretation of Information Capacity Theorem

Capacity is interpreted as the ability of a channel to convey information and is related to the noise characteristic of the channel. Shannon's capacity theorem [178] defines the fundamental limit on the rate of error-free transmission over a noisy communication channel. The information capacity of a channel is defined as the maximum of the *mutual information* between the channel input and the channel output over all distributions on the input that satisfy the power constraint [179, 180].

However, capacity is a theoretical ultimate transmission rate for reliable communication over a noisy channel. In practice, depending on the acceptable probability of error, there is a gap between the channel capacity and what is achievable by a practical coding and modulation scheme, called signal-to-noise ratio (SNR) gap, Γ, which is zero at theoretical capacity [181].

The information capacity of a continuous channel of bandwidth B Hz, perturbed by additive white Gaussian noise of power spectral density $N_0/2$ and limited in bandwidth to B, is given by

$$C = B \log_2 \left(1 + \frac{P}{N_0 B}\right), \tag{3.80}$$

where P is the average transmitted power. The above formula reveals the interplay among three key Parameters: channel bandwidth, average transmitted power, and noise power spectral density. While the dependence of the information capacity, C, on channel bandwidth, B, is linear, its dependence on SNR, P/N_0B, is logarithmic. Therefore, it will be easier to increase the information capacity of a communication channel by expanding its bandwidth rather than increasing the transmit power for a prescribed noise variance [180].

In a cognitive radio network, the communication channel is often shared between several transmitter–receiver pairs and information exchange between each pair interferes with the communication between the others. Such a channel is called an *interference channel* [182]. The capacity of interference channels is poorly understood even for simple cases. The set of all possible data rates achievable by all users is called the *rate region*. The sum-rate expression is a nonconvex function and finding the optimal power allocations for different users that guarantees the global maximum sum-rate is in general an NP-hard problem [183, 184].

Instead of solving the optimization problem globally, we settle for a suboptimal solution by viewing the problem as a noncooperative game [181]. The competing users try to maximize their data rates greedily by distributing their powers in a channel above the noise level but below a constant level determined by the permissible interference level (Figure 3.5). It is called the *waterfilling (pouring)* interpretation in

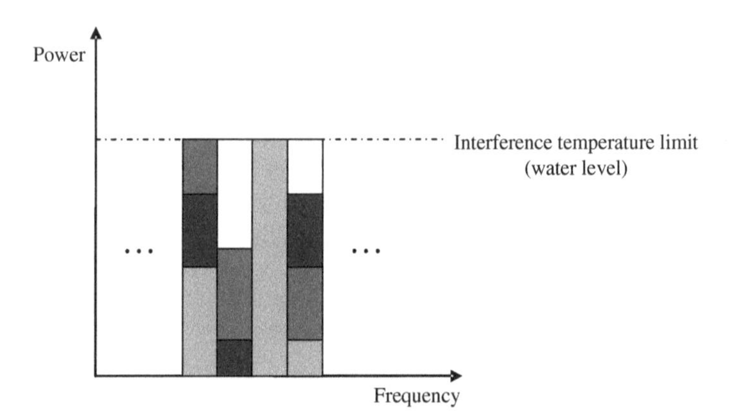

FIGURE 3.5 Waterfilling interpretation of the information-capacity theorem.

the sense that the process by which power is distributed is identical to the way in which water distributes itself in a vessel [180].

3.3.2 Iterative Waterfilling Algorithm (IWFA)

Finding a Nash equilibrium for the digital subscriber line (DSL) game was reformulated as a *nonlinear complementarity problem* (NCP) in [185]. In an NCP, the vector $\mathbf{x} \in \mathbb{R}^n$ should be found such that

$$\mathbf{x} \geq 0, \quad \mathbf{F}(\mathbf{x}) \geq 0, \quad \mathbf{x}^T\mathbf{F}(\mathbf{x}) \geq 0, \tag{3.81}$$

where $\mathbf{F}$ is a nonlinear mapping from $\mathbb{R}^n$ to $\mathbb{R}^n$. The problem will be a *linear complementarity problem* (LCP) if $\mathbf{F} = \mathbf{M}\mathbf{x} + \mathbf{q}$ for a matrix $\mathbf{M}$ and a vector $\mathbf{q}$ with appropriate dimensions [186]. In [187], the DSL game problem was reformulated as an LCP. Reformulation of the IWFA as an NCP and an LCP provides very interesting insights into this problem such as establishing the linear convergence under certain conditions on interference gains. Also, conditions on interference gains are obtained to guarantee convergence of the algorithm to a unique Nash-equilibrium point [167, 168, 187, 188]. However, the algorithm has some drawbacks:

- It is suboptimal.
- It is defenseless against clever selfish users that try to exploit dynamic changes or limited resources.

Moreover, regarding the dynamic nature of the cognitive radio environment and the speed of changes from one cycle to the next, the current transmit power values may not provide a good initial point for the next iteration. In this case, it may be better to start the iterative procedure from a randomly picked initial point in the new feasible set.

In what follows, the resource-allocation problem in cognitive radio networks is presented in the IWFA framework. While the predictive model can help for dealing with the appearance and disappearance of spectrum holes, robustification of the algorithm is proposed to address the issues related to unavoidable changes in the number of users and their mobility.

To proceed with the transmit-power control problem, assume that there are n active cognitive radio transmitter–receiver pairs in the region of interest and m subcarriers in an OFDM framework that could potentially be available for communication. Let PS denote the subset of subcarriers that are being used by primary users and some adjacent subcarriers that are used as guard bands, which cannot be assigned to cognitive radio users. Since spectral efficiency is the main goal of cognitive radio, the utility function chosen by each user to be maximized is the data rate. To be specific, user i solves the following optimization problem:

$$\max_{\mathbf{p}^i} f^i(\mathbf{p}^1, \ldots, \mathbf{p}^n) = \sum_{k=1}^{m} \log_2\left(1 + \frac{p_k^i}{I_k^i}\right)$$

$$\text{subject to } \sum_{k=1}^{m} p_k^i \le p_{\max}^i \tag{3.82}$$

$$p_k^i + I_k^i \le \mathrm{CAP}_k, \ \forall k \notin \mathrm{PS}$$

$$p_k^i = 0, \ \forall k \in \mathrm{PS}$$

$$p_k^i \ge 0.$$

This formulation is called *rate-adaptive waterfilling*, where the notion of waterfilling is attributed to information theory [179]. p_k^i denotes user i's transmit power on subcarrier k. The noise plus interference experienced by user i on subcarrier k because of transmissions due to other users is

$$I_k^i = \sigma_k^i + \sum_{j \neq i} \alpha_k^{ij} p_k^j. \tag{3.83}$$

Since cognitive radio is *receiver centric*, I_k^i is measured at receiver i. The positive parameter σ_k^i is the normalized background noise power at user i's receiver input on the kth subcarrier. The nonnegative parameter α_k^{ij} is the normalized interference gain from transmitter j to receiver i on subcarrier k and we have $\alpha_k^{ii} = 1$. Also, α_k^{ij} is proportional to $(d_{ii}/d_{ij})^r$, where d_{ij} is the distance from transmitter j to receiver i. Therefore, in general, $\alpha_k^{ij} \neq \alpha_k^{ji}$. If user i's receiver is closer to its transmitter compared to other active transmitters in the network, we will have $\alpha_k^{ij} \le 1$. $p_{\max}^i$ is user i's maximum power and CAP_k is the maximum allowable interference on subcarrier k. The first constraint states that the total transmit power of user i on all subcarriers should not exceed its maximum power (i.e., power budget). The second constraint set guarantees that the interference caused by all cognitive radio users on each subcarrier will be less than the maximum allowed interference on that subcarrier to guarantee an acceptable level of quality of service.

The waterfilling algorithm is implemented in a decentralized manner. In order to solve the optimization problem (3.82), user i does not need to know the value of p_j^k $\forall j \neq i$. The I_k^i defined in (3.83) is measured by user i's receiver rather than calculated, and therefore there is no need for information exchange between users. It is not even necessary for user i to know the number of other users in the network. Therefore, changing the number of the users in the network does not affect the complexity of the optimization problem that should be solved by each user. Hence, there is not any scaling problem [66]. The waterfilling game described in (3.82) is a *concave game* with *coupled constraints* for which an equilibrium point always exists [189].

The feasible set of the optimization problem that user i tries to solve, $\mathcal{K}^i$, depends on other users' decisions $\mathbf{p}^{-i}$, and, in effect, the joint feasible set of the entire game,

$\mathcal{K}$, does not have a Cartesian structure (i.e., $\mathcal{K} \subset \prod_{i=1}^{n} \mathcal{K}^i$). Thus, the equilibrium problem is called a *generalized Nash equilibrium* in the sense described in [190, 191]. Such a noncooperative n-person game with constraints can also be interpreted as an economic model with externalities [192]. In any event, the Nash equilibrium achieved by a decentralized resource-allocation approach may not be *Pareto optimal*. Hence, the equilibrium may be far from an optimal solution, which can be provided by a centralized approach [193]. In other words, in noncooperative games, phenomena such as *tragedy of the commons* usually lead to inefficient equilibria [194].

In order to improve the quality of the reached equilibrium in heterogeneous wireless networks, a hybrid approach based on the combination of network-centric (i.e., centralized) and terminal-centric (i.e., decentralized) radio resource management schemes was proposed in [195]. Alternative approaches such as coalitional games [196] and pricing [197] have also been proposed to address the issue of equilibrium inefficiency.

While the action of user i is denoted by its power vector $\mathbf{p}^i$, following the notation in the game theory literature, the joint actions of the other $n-1$ users are denoted by $\mathbf{p}^{-i}$. Three major types of adjustment schemes, $\mathcal{S}$, can be used by the users to update their actions [76]:

(i) Iterative waterfilling [188, 198, 199]: Users update their actions in a predetermined order:

$$\mathbf{p}^{-i}(\mathcal{S}_t) = [\mathbf{p}^1(t+1), \dots, \mathbf{p}^{i-1}(t+1), \mathbf{p}^{i+1}(t), \dots, \mathbf{p}^n(t)]. \quad (3.84)$$

(ii) Simultaneous iterative-waterfilling [167]: Users update their actions simultaneously regarding the most recent actions of the others:

$$\mathbf{p}^{-i}(\mathcal{S}_t) = \mathbf{p}^{-i}(t). \quad (3.85)$$

(iii) Asynchronous iterative-waterfilling [168]: It is an instance of an adjustment scheme that user i receives update information from user j at random times with delay:

$$\mathbf{p}^{-i}(\mathcal{S}_t) = [\mathbf{p}^1(\tau_t^{i,1}), \dots, \mathbf{p}^{i-1}(\tau_t^{i,i-1}), \mathbf{p}^{i+1}(\tau_t^{i,i+1}), \dots, \mathbf{p}^n(\tau_t^{i,n})], \quad (3.86)$$

where $\tau_t^{i,j}$ is an integer-valued random variable satisfying

$$\max(0, t-d) \leq \tau_t^{i,j} \leq t+1 \quad j \neq i \quad i,j \in \mathbb{N}, \quad (3.87)$$

which means that the delay does not exceed d time units.

Due to lack of central scheduling and difficulty of synchronization between different users in a cognitive radio network, the asynchronous adjustment scheme is more realistic than the other two.

3.3.3 IWFA as a Multistage Optimization Problem in Light of System Uncertainties

Since a cognitive radio network is a hybrid dynamic system, policies are defined on the event space as well as on the state space and therefore, each user needs to solve the corresponding optimization problem in two stages based on events and states.

- **Event-Based Optimization**: A set of state transitions is called an event. Events determine the dimension of the state space. When primary users stop communication, they release subbands, which can be used by cognitive radios. This event increases the dimension of the optimization problems that are solved by secondary users. On the other hand, when primary users start communication, they occupy subbands. This event decreases the dimension of the optimization problems that are solved by secondary users. Each user's dynamic spectrum manager chooses a set of appropriate channels for communication. Finding the optimal set of channels for each user is equivalent to the well-known graph coloring problem in graph theory [46]. In [200], a novel self-organizing spectrum management scheme is proposed, which uses *Hebbian learning* [48, 132] and solves the problem in a decentralized manner. This way, cognitive radios will be able to learn communication patterns of the primary users and build a predictive model, which determines the control horizon for the transmit-power controller. In the time intervals between such events, the state dimension of each user remains unchanged and state-based optimization is performed to find the optimal transmit power vector.
- **State-Based Optimization**: In the time intervals in which the available spectrum holes are fixed, the cognitive radio environment still has a dynamic nature, secondary users move all the time, they can leave the network and new users can join the network in a stochastic manner. Because of these activities, the interference plus noise term (3.83) in the objective function and the second constraint set of the optimization problem (3.82) are both time varying; the IWFA therefore assumes the form of an optimization problem under uncertainty. As mentioned in Sections 1.8 and 1.9, stochastic and robust optimization can be employed to deal with the uncertainty caused by joining and leaving of other cognitive radios as well as their mobility. After discussing the pros and cons of these two approaches, it was concluded that the robust optimization is a more reasonable approach, hence the material that follows.

3.3.4 Robust IWFA

Because of different sources of uncertainty, the noise plus interference term is the summation of two components: a nominal term, $\bar{I}$, and a perturbation term, ΔI, as

$$I_k^i = \bar{I}_k^i + \Delta I_k^i. \tag{3.88}$$

In the following, the objective functions for both stochastic and robust versions of the optimization problem (3.82) are presented.

If there is good knowledge about the probability distribution of the uncertainty term, ΔI, the IWFA problem (3.82) can be formulated as a stochastic optimization problem with the following objective function:

$$\max_{\mathbf{p}^i} \left[\mathbb{E}_{\Delta \mathbf{I}^i} \sum_{k=1}^{n} \log_2 \left(1 + \frac{p_k^i}{\bar{I}_k^i + \Delta I_k^i} \right) \right], \tag{3.89}$$

where $\mathbb{E}$ denotes the statistical expectation operator and

$$\Delta \mathbf{I}^i = [\Delta I_1^i, \ldots, \Delta I_m^i]^T. \tag{3.90}$$

The formulation of IWFA as a robust game in the sense described in [201] is basically a *max–min* problem in which each user tries to maximize its own utility while the environment and the other users are trying to minimize that user's utility [77, 93]. The worst-case interference scenarios have been studied for DSL in [202]. Considering an ellipsoidal uncertainty set, the IWFA problem (3.82) can be formulated as the following robust optimization problem.

$$\max_{\mathbf{p}^i} \left[\min_{\|\Delta \mathbf{I}^i\| \leq \varepsilon} \sum_{k=1}^{m} \log_2 \left(1 + \frac{p_k^i}{\bar{I}_k^i + \Delta I_k^i} \right) \right] \tag{3.91}$$

$$\begin{aligned} \text{subject to} \sum_{k=1}^{n} p_k^i &\leq p_{\max}^i \\ \max_{\|\Delta \mathbf{I}^i\| \leq \varepsilon} (p_k^i + \bar{I}_k^i + \Delta I_k^i) &\leq \text{CAP}_k, \ \forall k \notin \text{PS} \\ p_k^i &= 0, \ \forall k \in \text{PS} \\ p_k^i &\geq 0. \end{aligned}$$

A larger ε accounts for larger perturbations and the second set of constraints guarantees that the permissible interference power level will not be violated for any perturbation from the considered uncertainty set.

3.3.5 The Price of Robustness

In addition to conservatism, there is yet another price to be paid for achieving robustness. Although the IWFA problem (3.82) is a convex optimization problem, appearance of the perturbation term, ΔI, in the denominator of signal-to-interference plus noise ratio (SINR) in the objective function of the robust IWFA problem (3.91), makes it a nonconvex optimization problem. A robust optimization technique is proposed in [203] for solving nonconvex and simulation-based problems. The proposed method is based on the assumption that the cost and constraints as well as their gradient values are available. The required values can even be provided by numerical simulation subroutines. It operates directly on the surface of the objective function, and therefore does not assume any specific structure for the problem. In this method, the robust optimization problem is solved in two steps, which are applied repeatedly in order to achieve better robust designs.

- *Neighborhood search*: The algorithm evaluates the worst outcomes of a decision by obtaining knowledge of the cost surface in a neighborhood of that specific design.
- *Robust local move*: The algorithm excludes neighbors with high costs and picks an updated design with lower estimated worst-case cost. Therefore, the decision is adjusted in order to counterbalance the undesirable outcomes.

Linearity of constraints of the robust optimization problem (3.91), especially the second set of constraints that involves the perturbation terms, improves the efficiency of the algorithm.

3.3.6 Robust IWFA versus Classic IWFA

Simulation results are now presented to support theoretical underpinnings of this section. It is assumed that the cognitive-radio transceivers are distributed randomly in the region of interest with uniform distribution (Figure 3.6); this assumption is intuitively satisfying. Similar to [185, 187], the normalized interference gains α_k^{ij} are chosen randomly from the interval $(0, 1/(n-1))$ with uniform distribution, which are less than $1/(n-1)$, in order to guarantee that the tone matrices will be strictly diagonally dominant. Thus, the corresponding matrix $-\mathbf{M}$ will be Hurwitz. The ambient noise is assumed to be zero-mean Gaussian and the variance of the noise experienced by each user on each subcarrier is chosen from the interval $(0, 0.1/(n-1))$ with uniform distribution. The power budgets $p_{\max}^i$ are chosen randomly from the interval $(m/2, m)$ with uniform distribution too. For scenarios that consider the time-varying delay in the control loops, delays are chosen randomly. As shown in Figure 3.7, user mobility changes the communication path partially or completely, which, in turn, changes the interference gains and matrix $\mathbf{M}$. The same thing happens when new users join the network or old users leave the network.

3.3.6.1 Stochastic Variability in the Network Configuration

The transmit power control problem in a cognitive radio network using the classic IWFA and its robust version were discussed previously. In a cognitive radio network, when a spectrum hole disappears, users may have to seek or else increase their transmit powers in other spectrum holes and this increases the interference. Also, when new users join the network, current users in the network experience more interference. Therefore, the joining of new users or the disappearance of spectrum holes makes the interference condition worse. Also, the cross-interference between users is time varying because of mobility of the users. Results related to two typical but extreme

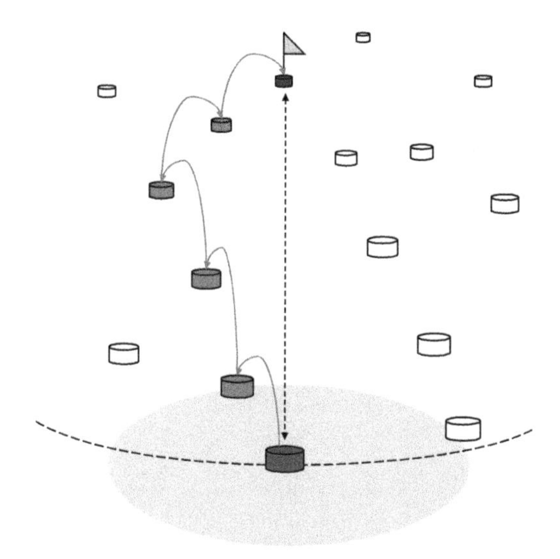

FIGURE 3.6 Multihop communication path between a source node and a destination node.

scenarios are presented here to show superiority of the robust IWFA (3.91) over the classic IWFA (3.82) in dealing with the above issues.

The first scenario addresses a network with $n = 5$ nodes and $m = 2$ available subcarriers, and all of the users simultaneously update their transmit powers using the interference measurements from the previous time step. As mentioned previously, the asynchronous adjustment scheme is the most realistic one when network simplicity is at a premium. However, here simultaneous adjustment was employed to implement extreme cases, which emphasizes the practical effectiveness of robust IWFA and its superiority over the classic IWFA. At the fourth time step, two new users join the network, which increases the power level of interference. The interference gains are also changed randomly at different time instants to consider mobility of the users. Figures 3.8 and 3.9 show the transmit power of three users (users 1, 4, and 7) on two different subcarriers for the classic IWFA and robust IWFA, respectively. At the second subcarrier, the classic IWFA is not able to reach an equilibrium. Data rates achieved by the chosen users are shown too. Also, the total data rate in the network is plotted against time, which is a measure of spectral efficiency. Although the average sum rate achieved by the classic IWFA is close to the average sum rate of the robust IWFA, it fluctuates and in some time instants the data rate is very low, which indicates lack of spectrally efficient communication. Although, the oscillation occurs mainly because of using simultaneous update scheme, it also highlights practical effectiveness of the robust IWFA.

In the second scenario, a network with $n = 5$ nodes and $m = 4$ available subcarriers is considered. Again at the fourth time step, two new users join the network but at the

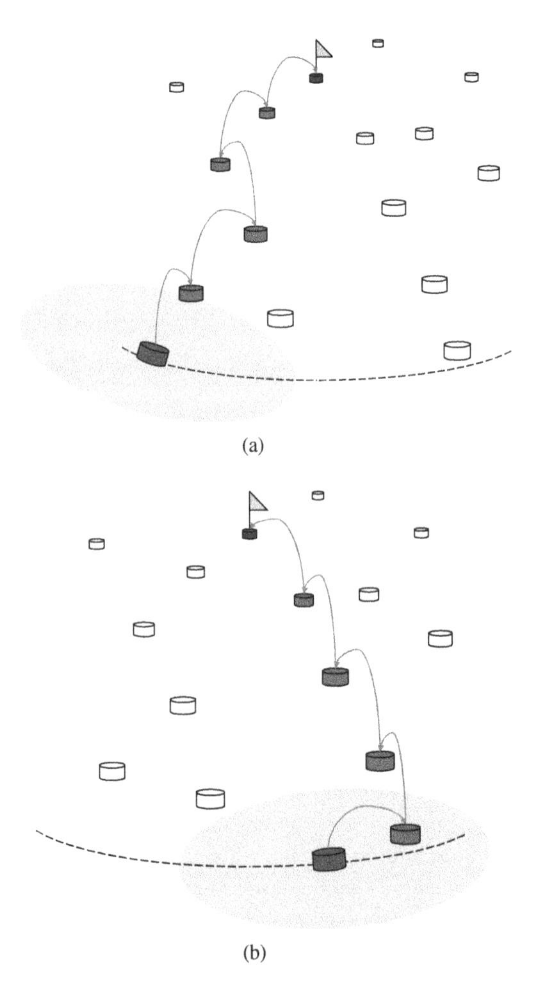

FIGURE 3.7 Effect of user mobility on the communication path: (a) partially changed, (b) completely changed with respect to the path shown in Figure 3.6.

eighth time step the third subcarrier is not available anymore (i.e., a spectrum hole disappears). Results are shown in Figures 3.10 and 3.11, which, again show superiority of the robust IWFA. For classic IWFA, immediately after the disappearance of the third subcarrier, power in the fourth subcarrier starts to oscillate and after changing the interference gains randomly, we observe the same behavior in other subcarriers. In contrast to the robust IWFA, the classic IWFA fails again to achieve an equilibrium.

3.3.6.2 Delay As mentioned previously, sporadic feedback introduces a time-varying delay in the transmit power control loop, which causes different users to update their transmit powers based on outdated statistics. For instance, when the network configuration and therefore interference pattern changes, some users receive

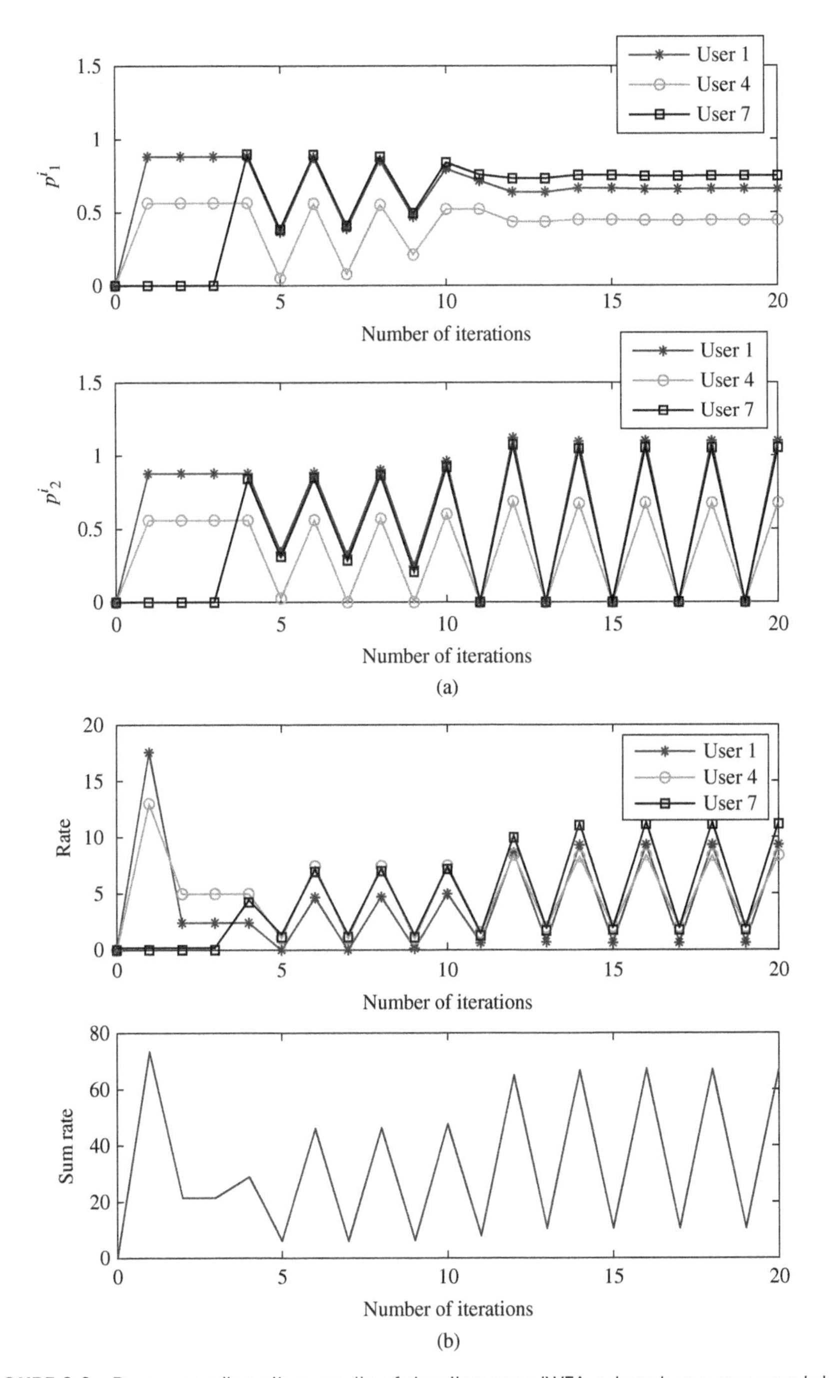

FIGURE 3.8 Resource-allocation results of simultaneous IWFA, when two new users join a network of five users and interference gains are changed randomly due to mobility of the users: (a) transmit powers of three users on two subcarriers, (b) data rates of three users and the total data rate in the network. Source: Setoodeh and Haykin (2009) [66]. Reproduced with the permission of IEEE.

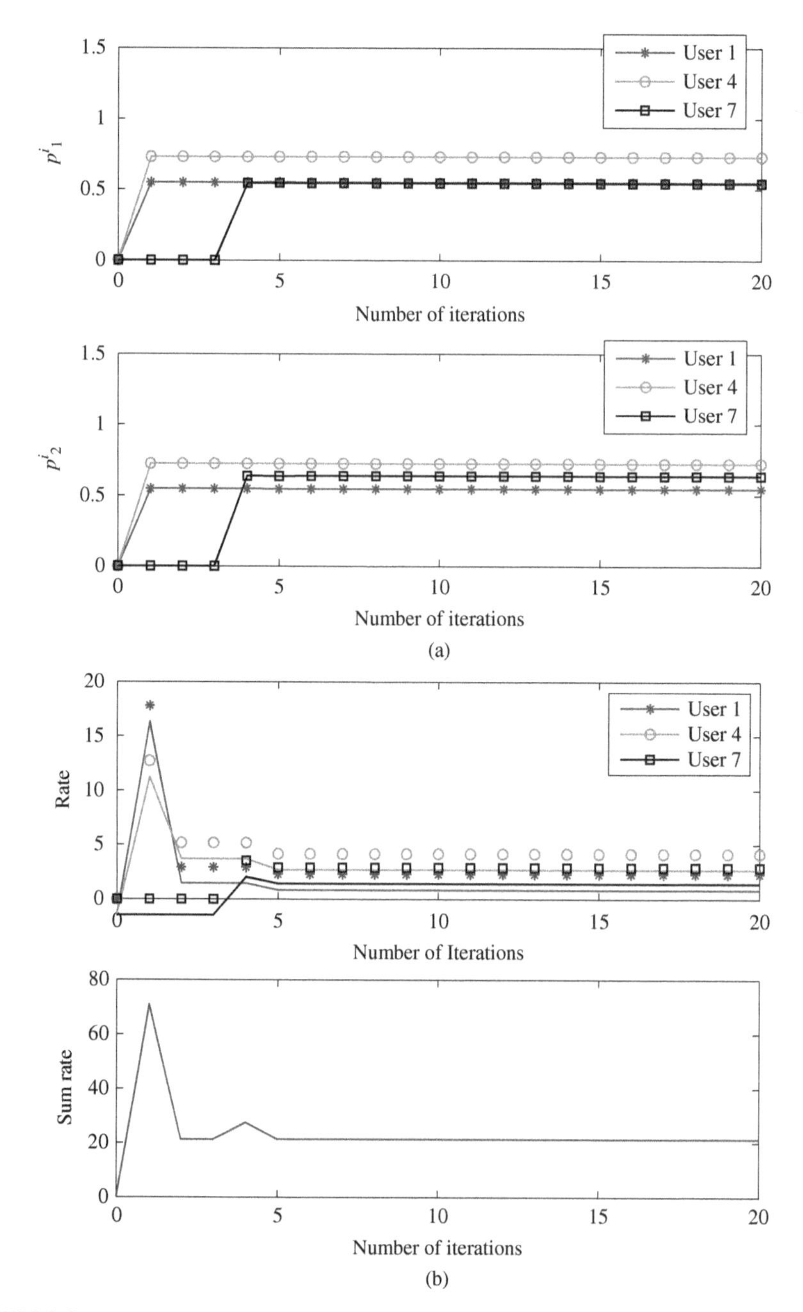

FIGURE 3.9 Resource-allocation results of simultaneous robust IWFA, when two new users join a network of five users and interference gains are changed randomly due to mobility of the users: (a) transmit powers of three users on two subcarriers, (b) data rates of three users and the total data rate in the network. Source: Setoodeh and Haykin (2009) [66]. Reproduced with the permission of IEEE.

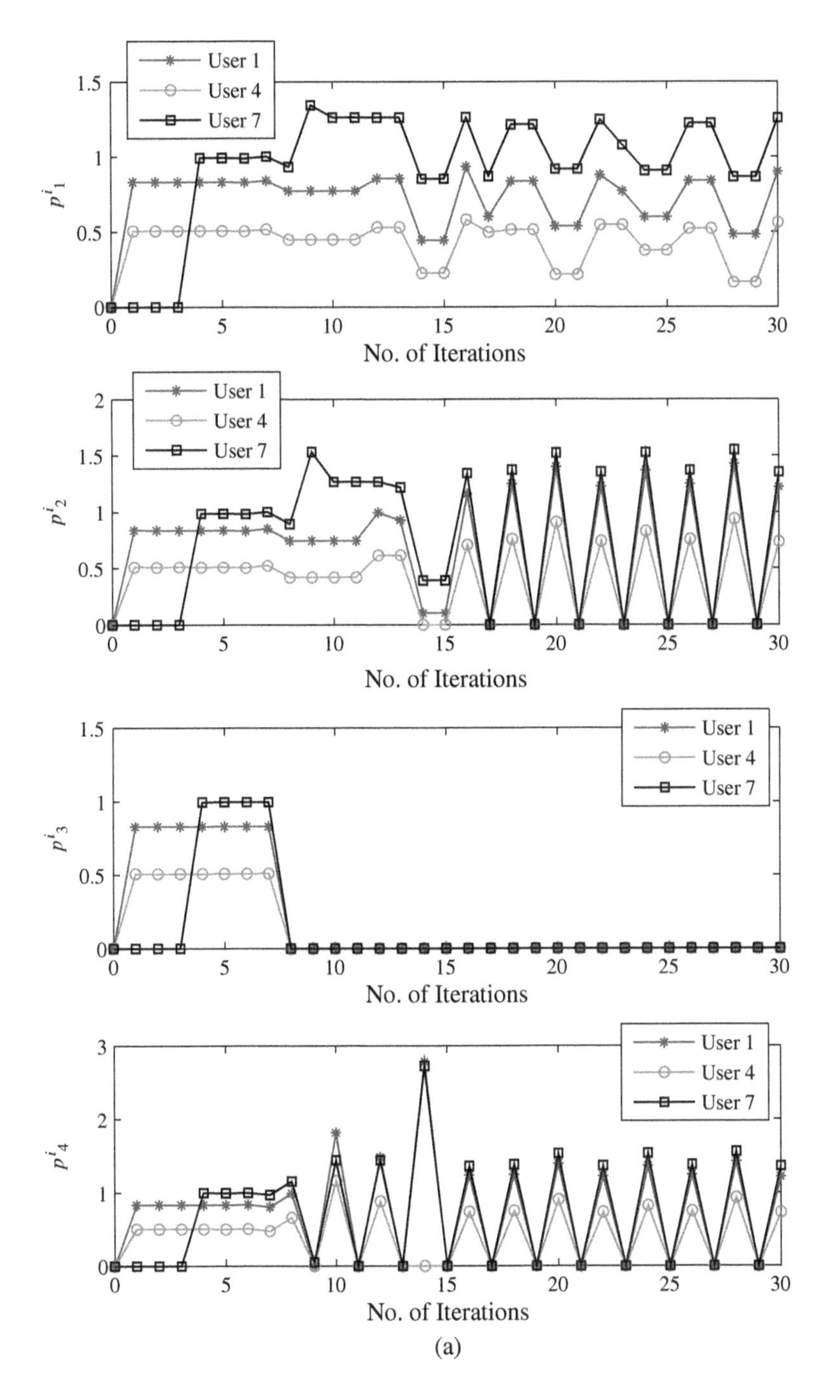

FIGURE 3.10 Resource-allocation results of simultaneous IWFA, when two new users join a network of five users, a subcarrier disappears, and interference gains are changed randomly due to mobility of the users: (a) transmit powers of three users on four subcarriers, (b) data rates of three users and the total data rate in the network. Source: Setoodeh and Haykin (2009) [66]. Reproduced with the permission of IEEE.

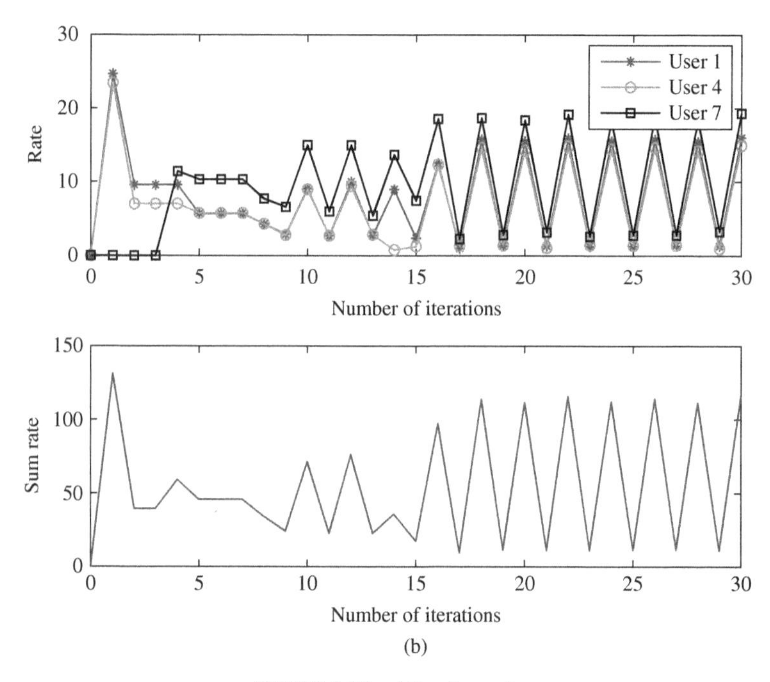

FIGURE 3.10 *(Continued)*

the related information after a delay. If the interference on a subcarrier increases and the transmitter is not informed immediately, it will not reduce its transmit power and it may violate the permissible interference power level for a while until it receives updated statistics of the interference in the forward channel. Similarly, this may happen to some users that update their transmit powers at lower rates compared to others. In the third scenario, a new user joins a network of three users, who are competing for utilizing two subcarriers. Each user's transmitter receives statistics of the interference plus noise with a time-varying delay. Figure 3.12(a) shows the randomly chosen time-varying delays introduced by each user's feedback channel. Sum of transmit power and interference plus noise at the second subcarrier at the receiver of each user is plotted in Figure 3.12(b) and (c) for classic IWFA and robust IWFA, respectively. Dashed lines show the limit imposed by the permissible interference power level. Although the classic IWFA is less conservative, it is not as successful as the robust IWFA in preventing violations of the permissible interference power level. Similar results are obtained when users update their transmit powers with different frequencies.

These small-scale problems were designed and typical results were presented to compare the performance of classic IWFA and robust IWFA. In some extreme cases, because of occurrence of discrete events such as the appearance and disappearance of spectrum holes and users, the IWFA cannot achieve an equilibrium solution and calculated results oscillate in subsequent time steps especially if we use the simultaneous

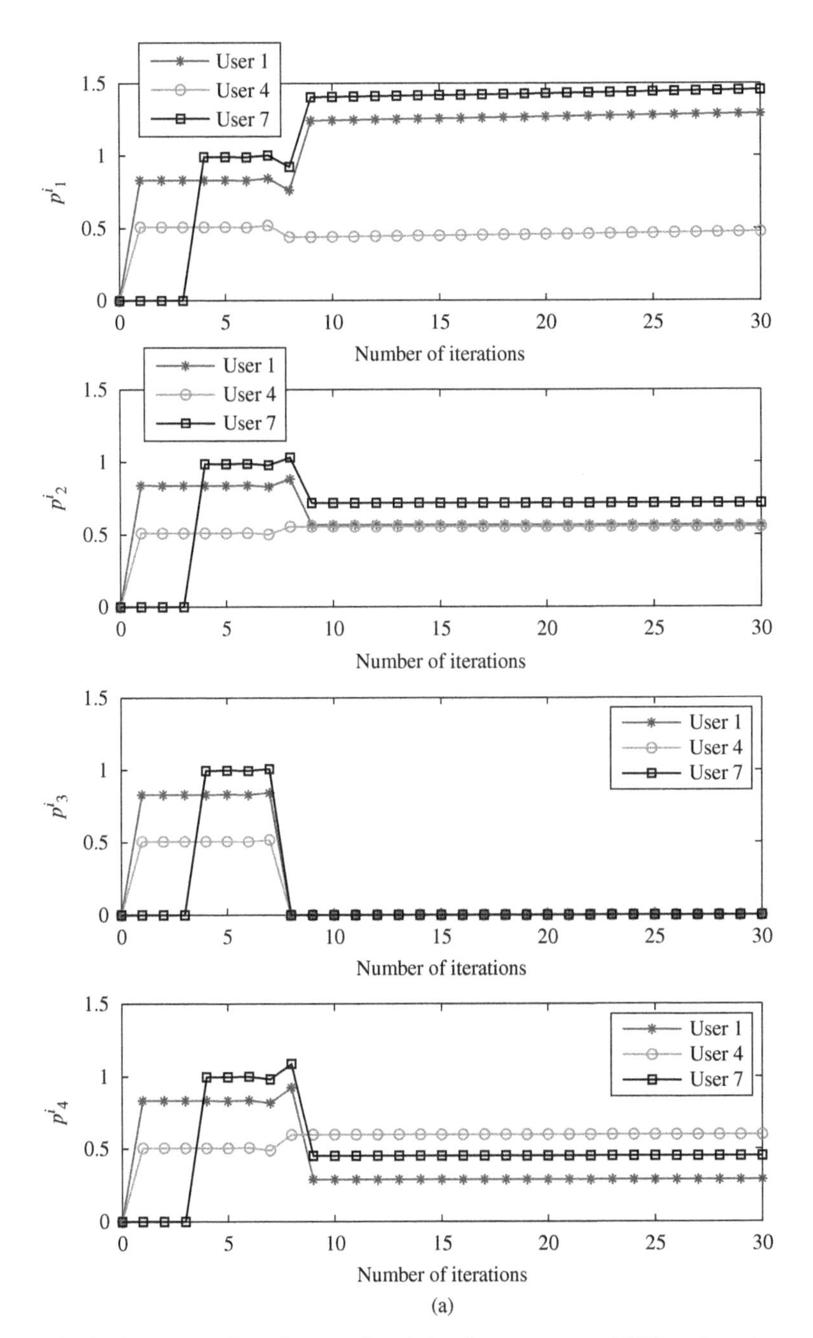

FIGURE 3.11 Resource-allocation results of simultaneous robust IWFA, when two new users join a network of five users, a subcarrier disappears, and interference gains are changed randomly due to mobility of the users: (a) transmit powers of three users on four subcarriers, (b) data rates of three users and the total data rate in the network. Source: Setoodeh and Haykin (2009) [66]. Reproduced with the permission of IEEE.

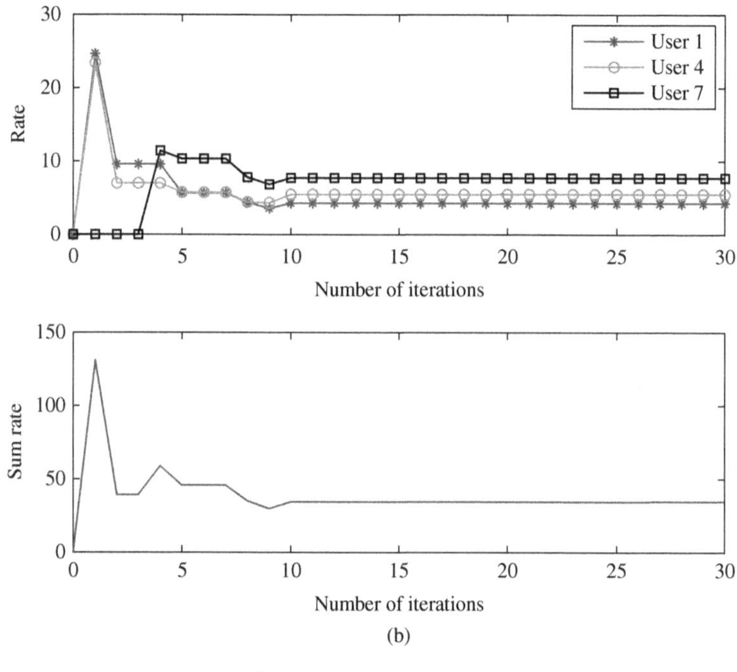

FIGURE 3.11 *(Continued)*

update scheme. This confirms the point that in a hybrid dynamic system such as a cognitive radio network in which switching happens between different subsystems even if all of them are stable, the whole system may become unstable because of switching between the subsystems. In the presented cases, the robust IWFA was able to achieve an equilibrium solution. Also, when some users update their transmit powers with lower frequencies or use outdated information, the robust IWFA can prevent violating the permissible interference power level. Classic IWFA lacks this ability although it achieves superior data rates.

To summarize, simulation results were presented to compare the performance of classic IWFA versus robust IWFA. Toy scenarios were considered in order to develop insight. Results show superiority of the robust IWFA over the classic IWFA in dealing with different practical issues in a cognitive radio environment, which is achieved by putting up with a reduction in achievable data rate for reliable performance. As mentioned previously, IWFA is defenseless against malicious users that do not follow the rules and do not decrease their transmit power, when the interference level is high. Such users can exploit the limited resources and achieve higher data rates compared to well-behaved users. In effect, therefore, a malicious user may act like a jammer and have the same effect on the network that disappearance of a spectrum hole has. Hence, other users' power vectors may fluctuate and the network may not be able to reach an equilibrium. Similar results on oscillation of transmit powers and, therefore, data rates in the presence of a jammer were reported in [204]. In this situation, robust

IWFA enables the well-behaved users to reach an equilibrium with possibly lower data rates.

3.4 INFORMATION VALUE

The key role that even limited or finite-rate feedback can play in wireless communications has been explored by many researchers [205, 206]. Employing limited feedback for channel adaptive signaling in wireless communication systems allows the transmitter to take account of channel condition for interference avoidance and, therefore, significantly improves the performance. A cognitive radio transceiver takes the form of a closed-loop control system, where the feedback channel is used for two purposes: first, establishing a communication channel between transmitter and receiver, and then channel adaptive signaling. Hence, the information on the available spectrum holes has a much more priority than the information on the forward channel condition because the operation of cognitive radio totally depends on the former. However,

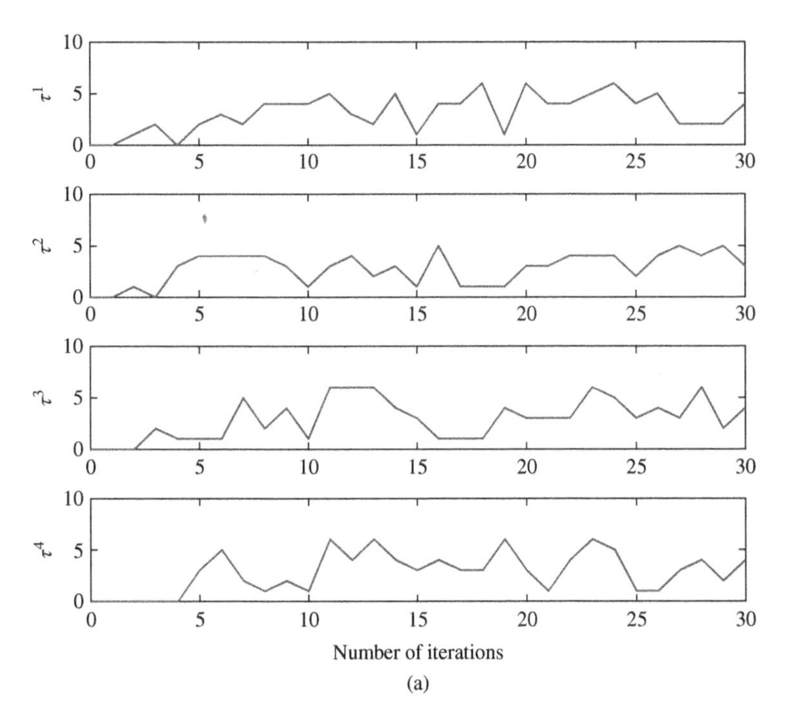

FIGURE 3.12 Resource-allocation results of IWFA, when interference gains change randomly with time and users use outdated information to update their transmit powers: (a) time-varying delays introduced by each user's feedback channel. Sum of transmit power and interference plus noise for four users achieved by (b) classic IWFA and (c) robust IWFA. Dashed lines show the limit imposed by the permissible interference power level. Source: Setoodeh and Haykin (2009) [66]. Reproduced with the permission of IEEE.

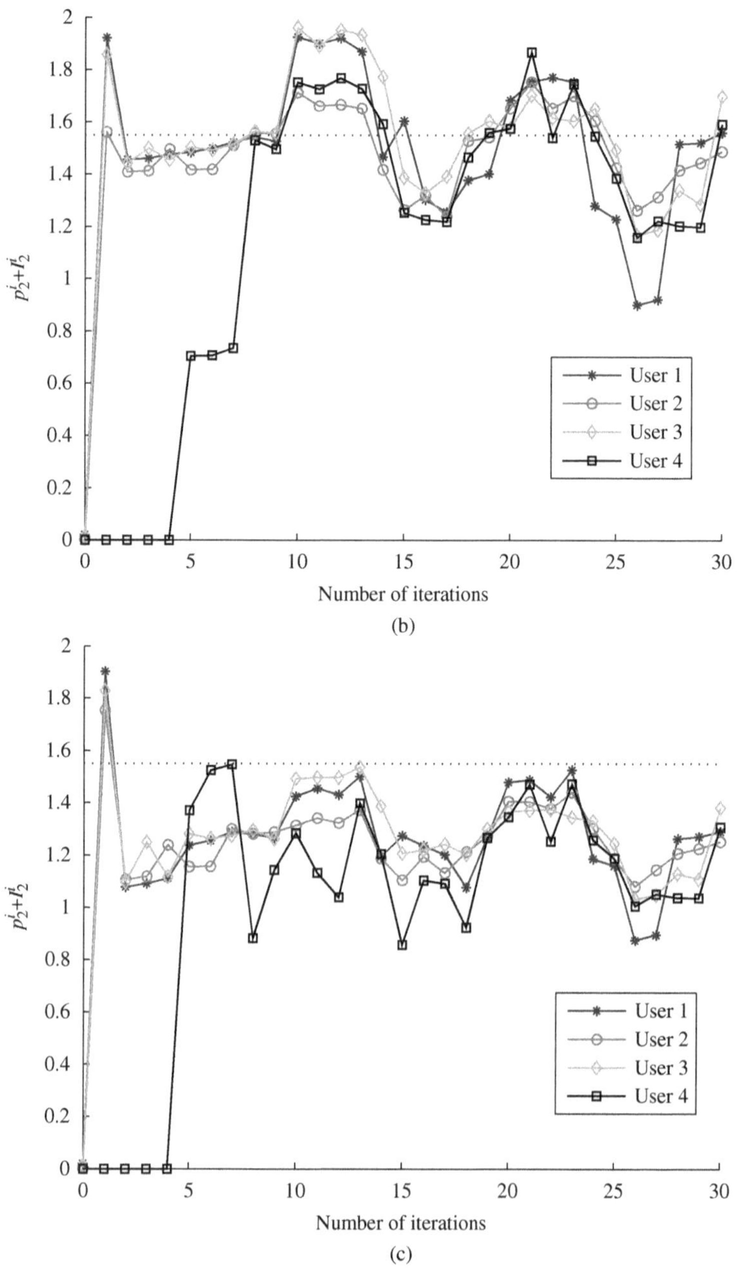

(b)

(c)

FIGURE 3.12 *(Continued)*

gathering and processing information comes at a cost that must be taken into account in the process of decision-making.

In classical terms, Shannon's celebrated information theory was originally developed to mathematically formalize the transmission of signals through a communication channel [207]. The theory provides a quantitative measure of the amount of information, which depends only on the probabilistic structure of the communication channel under study. Information theory has found diverse applications beyond just transmission and compression of data.

From a control or decision-making point of view, the probabilistic nature of uncertainties as well as their (economic) impacts on the decision-maker must be taken into account and a theory that only deals with probabilities of outcomes may not completely describe the importance of uncertainty to the decision-maker. When it comes to allocation of computational resources for information processing, the *value of information* is of critical importance [208].

Therefore, a functional definition of information would be necessary; hence the notion of *control information*, which is defined in as follows [209]:

> Control information is the capacity (*know-how*) to *control the acquisition, disposition, and utilization* of matter/energy in purposive (cybernetic) processes.

Due to the information gathering cost and constraints on computational time and power, similar to the idea of limited feedback in traditional wireless communication systems, a limited informational bandwidth can be used for cognitive radios. This way, for making decisions and taking actions, the informational costs or the required informational bandwidth in the form of an *information-to-go* function can be considered in addition to the standard *value-to-go* function in the Bellman dynamic programming formulation [210]. This line of thinking paves the way for information-theoretic analysis of cognitive radio.

3.5 CONCLUDING REMARKS

A cognitive radio transceiver is built around a perception–action cycle with the perceptual and the executive parts being apart. Radio-scene analyzer in the receiver plays the role of the perceptual part, and dynamic-spectrum manager and transmit-power controller jointly play the role of the executive part. Memory plays a key role in the dynamic-spectrum manager and adding memory will significantly improve its performance. Scanning the electromagnetic spectrum consumes time, energy, and computational resources. Perceptual attention addresses this issue by focusing on specific portions of the spectrum. Therefore, four pillars of the Fuster's paradigm of cognition have been taken into account in designing a cognitive radio transceiver.

4

COGNITIVE RADIO NETWORKS

The infrastructures developed for production, distribution, and consumption of goods as well as providing services are referred to as supply chains [18]. All involved parties are decision-makers that follow their own interests in a way to maximize their profit and/or minimize risk. It follows therefore that decentralized decision-making is an inherent characteristic of supply chains. In a supply chain, the interactions between decision-makers must provide a sustainable flow of products or services and at the same time, consumers must be satisfied in terms of the price they pay for a particular product or service. In the context of cognitive radio networks, we may therefore make the following statement:

> A supply chain is a complex system whose dynamic behavior is affected by both competition and cooperation.

In effect, a holistic approach, which is systemwide and network based, would be required to study supply chains. This chapter builds on the framework, which was proposed in [37].

4.1 COGNITIVE RADIO NETWORKS VIEWED AS SPECTRUM-SUPPLY CHAIN NETWORKS

As mentioned previously in the introductory chapter, the combination of legacy and cognitive wireless worlds in a specific region can be viewed as a spectrum-supply chain network, in which the legacy owner and secondary users, respectively, play the role of suppliers and consumers of spectrum [59]. Figure 4.1 depicts a typical

Fundamentals of Cognitive Radio, First Edition. Peyman Setoodeh and Simon Haykin.

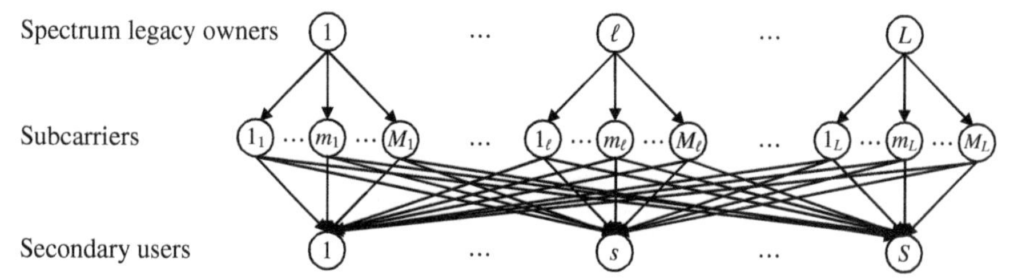

FIGURE 4.1 The spectrum-supply chain network in its basic form with two tiers: legacy owners and secondary users. In each tier of the network, a noncooperative game is played among the peers. In a market-driven regime, legacy owners compete against each other to gain more benefit from leasing their underutilized subbands, and secondary users compete against each other to get a better share from unlicensed bands and a share of the licensed bands at a lower price if needed. In an open-access regime, only one game is played among secondary users to get a better share from unlicensed bands as well as the idle licensed subbands of legacy owners. Source: Haykin and Setoodeh (2015) [37]. Reproduced with the permission of IEEE.

spectrum-supply chain network for S secondary users and L spectrum legacy owners, each of which owns spectrum subbands including M_ℓ channels ($\ell = 1, \ldots, L$) and provides service to a number of primary users. The spectrum-supply chain of Figure 4.1 provides a framework for holistic systemwide modeling and analysis of cognitive radio networks. The links between each channel and secondary users show that a specific channel can potentially be used by all secondary users. These channels can be considered as subcarriers in an OFDM scheme.

In the open-access regime, a game is played between secondary users for obtaining a better share from the available resources. Since secondary users compete against each other for limited resources, their *perception–action cycle* plays a critical role for their survival. Here, secondary users indirectly interact through spectrum holes, whose availability depends on the primary users' activity patterns [66, 211]. Thus, Figure 4.1 does apply to the open-access regime of cognitive radio networks.

In the market-driven regime, on the other hand, another group of decision-makers, namely, *brokers* may be part and parcel of the spectrum-supply chain [212]. To this end, Figure 4.1 is extended to a three-tier structure as depicted in Figure 4.2. Here, three different games are played in each tier among peer decision-makers that compete against each other while interacting with decision-makers in other tiers as well [59]. In either one of these two regimes, the large number of *heterogeneous* elements in the spectrum-supply chain network that interact with each other indirectly through the limited resources makes the cognitive radio network a *complex dynamic system* or a system of systems [211]. Moreover, dynamic behavior of the network is shaped by the phenomena of *stigmergy*, the essence of which is defined as follows [213]:

> The indirect communication is mediated by modifications of the environment.

In this kind of environment, different degrees of coupling between different decision-makers of one tier or between decision-makers from different tiers influence their chosen policies. Change of policies affects the interaction between the

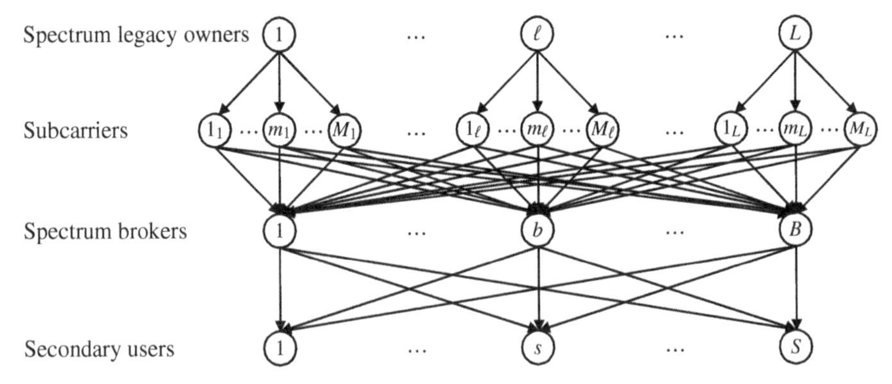

FIGURE 4.2 The extended spectrum-supply chain network for market-driven regime with three tiers: legacy owners, spectrum brokers, and secondary users. In each tier of the network, a noncooperative game is played among the peers. Legacy owners compete against each other to gain more benefit from leasing their underutilized subbands. Spectrum brokers compete against each other to maximize their profit by buying the right of using underutilized licensed subbands from legacy owners at a lower price and selling it to secondary users at a higher price. Secondary users compete against each other to get a better share from unlicensed bands and a share of the licensed bands at a lower price if needed. Source: Haykin and Setoodeh (2015) [37]. Reproduced with the permission of IEEE.

decision-makers and alters the degrees of coupling between them. In other words, both *upward and downward causations* play key roles in the network and lead to positive or negative *emergent behavior* [211].

There are two different viewpoints on *emergence*, as described in [214]

(i) In *philosophy of mind*, emergence refers to *seemingly irreducible* phenomena.
(ii) In *cognitive science*, emergence refers to phenomena that are *not explicitly programmed*.

Nevertheless, these two viewpoints have a lot in common with each other. Behavior of a complex system or network is determined by the behavior of its subsystems or components as well as the interactions between them. Often, such interactions are too important and cannot be ignored. Therefore, a reductionist point of view that is able to explain the behavior of components cannot explain the global behavior of the system, which, in turn, means that both of the mentioned viewpoints emphasize the fact that the whole is more important than its components acting individually, which is intuitively satisfying. While the first viewpoint states that, due to interactions, the global behavior is not reducible to local behaviors, the second viewpoint states that interactions will lead to global behavior that is not explicitly programmed in the components. That is why such irreducible or not explicitly programmed behaviors are said to be *emergent*.

Since the global behavior of the network cannot be reduced to the local behaviors of different elements, taking an approach to build a dynamic model that provides a global description of the network behavior is of critical importance. It is impractical to perform experiments with large decentralized wireless networks with hundreds or thousands of nodes in order to understand their global behavior. Therefore, the importance of analytical approaches is highlighted even more [79]. Such models enable us to predict the future and, based on the obtained knowledge, engineer it to improve network robustness against potential disruptions [211].

Recognizing that a cognitive radio network is a complex dynamic system formed by a group of interacting subsystems (i.e., cognitive radio users) and activities of the primary users determine the available subbands for secondary usage, it follows therefore that role of the primary users in a cognitive radio network can be interpreted as the role of a high-level-network controller that decides which resources can be used by cognitive radio users. Hence, the resource-allocation algorithm used by each cognitive radio user determines the share of that user from the available resources. The resource-allocation algorithms play the role of local controllers that control the corresponding subsystems in a decentralized manner. Thus, the relationship between the two wireless worlds resembles the relationship between landlords on the one hand and tenants on the other hand, in which the spectrum plays the role of the property. Regarding this relationship, the *decentralized hierarchical control* structure described in [215] may be considered for a cognitive radio network as depicted in Figure 4.3.

The hierarchical structure shown in Figure 4.3 is reminiscent of *Stackelberg* games. In this kind of games, the roles of players are not symmetric and the powerful players that have the ability to dictate their strategies are known as *leaders*. Other

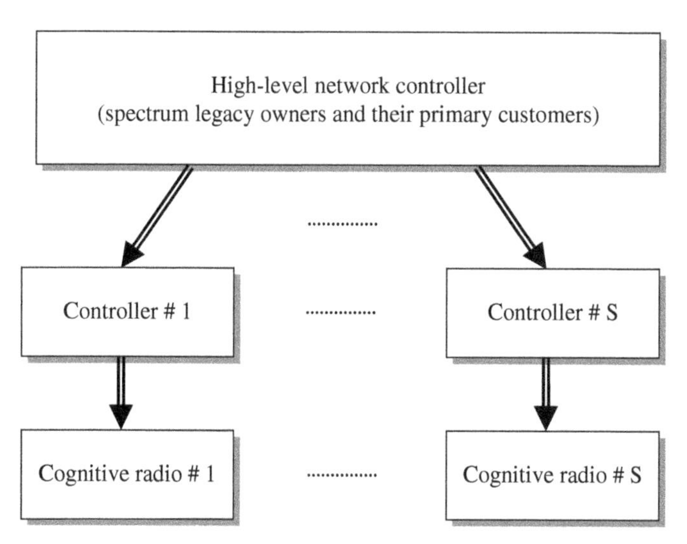

FIGURE 4.3 Decentralized hierarchical control structure in a cognitive radio network. Licensed bands are occupied and released according to the communication activities of primary users. These activities, which are discrete events, may be interpreted as the actions of a high-level network controller. On the other hand, the resource-allocation algorithms, which are employed by secondary users, may be viewed as local controllers. These local controllers are two-level controllers that handle channel assignment and transmit-power adjustment in a hierarchical manner. Source: Haykin and Setoodeh (2015) [37]. Reproduced with the permission of IEEE.

players known as *followers* must take actions by rationally reacting to leaders' decisions. This hierarchical decision-making process leads to a hierarchical equilibrium solution concept [76]. It is apparent from Figure 4.1 that legacy owners and their primary customers are able to enforce their strategies on secondary users. In effect therefore, the leader–follower relationship in a Stackelberg game is quite similar to our landlord–tenant interpretation in a spectrum-supply chain network. The transmit power allocation problem has been formulated as a two-tier Stackelberg game for both cellular [216] and cognitive radio networks [63]. Extensions to multilevels of hierarchy in decision-making, with many leaders and followers similar to Figure 4.2 is also permissible [76].

Since the network's overall behavior is a combination of both continuous-time dynamics and discrete events, we may refer to it as a *hybrid dynamic system* (HDS) [66]. When primary users stop communicating, they release subbands and when they start to communicate, they occupy subbands. Therefore, appearance and disappearance of spectrum holes, which are discrete events, occur due to the activities of primary users. These activities are interpreted as actions of the high-level network controller (Figure 4.3). On the other hand, we find that transmit-power adjustment according to the actions of the resource-allocation algorithms employed by secondary users (i.e., local controllers of cognitive radios in Figure 4.3) is continuous. The network state vector is composed of the continuous transmit-power vectors of cognitive radios and its evolution over the course of time represents the continuous-time dynamics. It should be noted that actions of the high-level network controller that are discrete events, and actions of the local controllers that are continuous, are, respectively, associated with *slow* and *fast* dynamics of the network; such a scenario leads to the two-timescale behavior [211].

It follows therefore that the resource-allocation problem should be solved in two stages representing discrete events and continuous states. In other words, the local controllers in Figure 4.3 are themselves two-level controllers [217, 218] including a *field-level controller* and a *supervisory-level controller*. In the control hierarchy, the latter has a higher rank than the former. The corresponding two-level control scheme is shown in Figure 4.4. The supervisory-level controller is, in effect, an event-based controller that deals with the appearance and disappearance of spectrum holes. The radio-scene analyzer will then inform the supervisory-level controller if it detects a change in the status of the available spectrum holes. In such a case, the supervisory-level controller chooses a new set of available channels and calls for reconfiguration of the transmitter in order to adapt the transmitting parameters to the chosen set of channels. In the perception–action cycle of a cognitive radio (Figure 1.1), dynamic spectrum manager has the role of the supervisory-level controller. On the other hand, the field-level controller is a state-based controller that adjusts the transmit power over the set of available channels chosen by the supervisory-level controller according to the interference level in the radio environment. In the perception–action cycle of a cognitive radio (Figure 1.1), the transmit power controller has the role of the field-level controller.

A cognitive radio may build an internal model for the external world (i.e., the radio environment), which is used to predict the availability of certain subbands, the

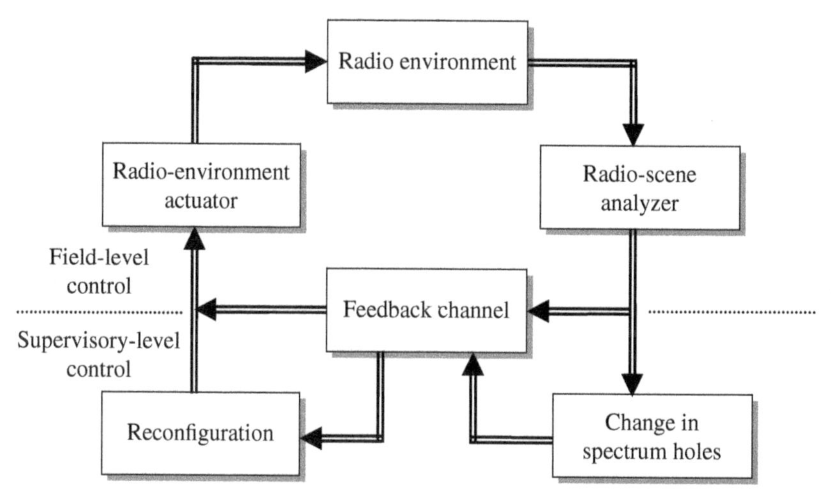

FIGURE 4.4 Two-level control scheme for cognitive radio. The controller is a hierarchical hybrid system. In the control hierarchy, the supervisory-level controller has a higher rank with respect to the field-level controller. The supervisory-level controller is an event-triggered controller and handles channel selection according to the primary users' communication patterns, which lead to appearance and disappearance of spectrum holes. In a cognitive radio, the dynamic spectrum manager plays the role of the supervisory-level controller. On the other hand, the field-level controller is a continuous state-based controller that adjusts the transmit power over the set of chosen channels. In a cognitive radio, the transmit-power controller plays the role of the field-level controller. Source: Haykin and Setoodeh (2015) [37]. Reproduced with the permission of IEEE.

duration of their availability, and approximate interference level in those subbands. These pieces of information will be of critical importance for providing a seamless communication in the dynamic wireless environment. Both the supervisory-level and the field-level controllers will benefit from a *predictive* model, which determines the control horizon for planning ahead.

4.2 OPEN-ACCESS COGNITIVE RADIO NETWORKS

The resource-allocation problem in cognitive radio networks in an open-access regime can be viewed as a noncooperative game [37]. To elaborate, suppose that there are n active cognitive radio transceiver pairs in the region of interest, and m subcarriers in an OFDM framework that could potentially be available for communication. Then, user i tries to solve the optimization problem in (3.82), which is reproduced here for convenience of presentation:

$$\max_{\mathbf{p}^i} \quad f^i(\mathbf{p}^1, \ldots, \mathbf{p}^n) = \sum_{k=1}^{m} \log_2 \left(1 + \frac{p_k^i}{I_k^i} \right)$$

$$\text{subject to}: \quad \sum_{k=1}^{m} p_k^i \leq p_{\max}^i \tag{4.1}$$
$$p_k^i + I_k^i \leq \mathrm{CAP}_k, \ \forall k \notin PS$$
$$p_k^i = 0, \ \forall k \in PS$$
$$p_k^i \geq 0.$$

To this end, let us concatenate the corresponding variables for all users as follows:

$$\mathbf{p} = \left[\mathbf{p}^i\right]_{i=1}^n = \begin{bmatrix} \begin{bmatrix} p_1^1 \\ \vdots \\ p_m^1 \end{bmatrix} \\ \vdots \\ \begin{bmatrix} p_1^n \\ \vdots \\ p_m^n \end{bmatrix} \end{bmatrix}, \tag{4.2}$$

$$\boldsymbol{\sigma} = \left[\boldsymbol{\sigma}^i\right]_{i=1}^n = \begin{bmatrix} \begin{bmatrix} \sigma_1^1 \\ \vdots \\ \sigma_m^1 \end{bmatrix} \\ \vdots \\ \begin{bmatrix} \sigma_1^n \\ \vdots \\ \sigma_m^n \end{bmatrix} \end{bmatrix}, \tag{4.3}$$

and

$$\mathbf{M} = \begin{bmatrix} \mathbf{M}^{11} & \cdots & \mathbf{M}^{1n} \\ \vdots & \cdots & \vdots \\ \mathbf{M}^{n1} & \cdots & \mathbf{M}^{nn} \end{bmatrix}, \tag{4.4}$$

where

$$\mathbf{M}^{ij} = \begin{bmatrix} \alpha_1^{ij} & \cdots & 0 \\ \vdots & \cdots & \vdots \\ 0 & \cdots & \alpha_m^{ij} \end{bmatrix}. \tag{4.5}$$

Characteristics of the joint feasible set, $\mathscr{K}$, and the affine mapping, namely,

$$\mathbf{F}(\mathbf{p}) = \boldsymbol{\sigma} + \mathbf{M}\mathbf{p} \tag{4.6}$$

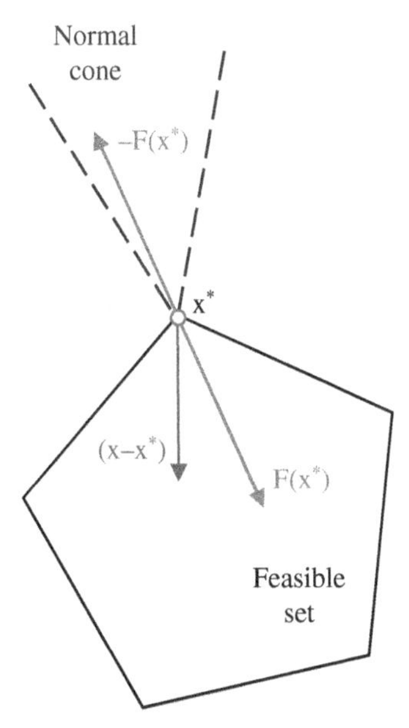

FIGURE 4.5 Geometric interpretation of variational inequalities. For all $\mathbf{x}$ in the feasible set, the equilibrium point denoted by $\mathbf{x}^*$ satisfies the inequality $(\mathbf{x} - \mathbf{x}^*)^T\mathbf{F}(\mathbf{x}^*) \geq 0$. Source: Haykin and Setoodeh (2015) [37]. Reproduced with the permission of IEEE.

reveal a great deal about network behavior. To be more precise, the waterfilling game can be formulated as an *affine variational inequality* (AVI) problem $\text{VI}(\mathcal{K}, \boldsymbol{\sigma} + \mathbf{Mp})$ or $\text{AVI}(\mathcal{K}, \boldsymbol{\sigma}, \mathbf{M})$ [66, 67, 187]. In other words, if $\forall \mathbf{p} \in \mathcal{K}$, the vector $\mathbf{p}^* \in \mathcal{K}$ satisfies the condition:

$$(\mathbf{p} - \mathbf{p}^*)^T(\boldsymbol{\sigma} + \mathbf{Mp}^*) \geq 0, \tag{4.7}$$

then, under this condition, $\mathbf{p}^*$ is an equilibrium point of the noncooperative game (see Appendix A for proof). The converse is true if the feasible set has a Cartesian structure [192]. Figure 4.5 illustrates a geometric interpretation of variational inequalities.

The AVI problem denoted by $\text{VI}(\mathcal{K}, \boldsymbol{\sigma} + \mathbf{Mp})$ may be interpreted as a robust optimization problem, in which the vector $\mathbf{p}$ is subject to uncertainty and known only to belong to $\mathcal{K}$ [219]. In what follows, it is shown that the AVI reformulation of the waterfilling game facilitates the study of the disequilibrium behavior and stability analysis of the cognitive radio network.

Although the components of the network may remain unchanged in complex and large-scale networks, the general behavior of the network can change dramatically over time. In particular, if the signal-to-interference plus noise ratio (SINR) of a communication link drops below a specified threshold for a relatively long time, then the connection between the transmitter and receiver will be lost. For this reason, in addition to the equilibrium resource allocation discussed previously, the transient behavior of the network deserves careful attention too [220]. Therefore, studying the equilibrium states in a dynamic framework by methods that provide information about the disequilibrium behavior of the system is of critical importance; this issue is the focus of the following subsection [66].

4.2.1 Network Dynamics

Theory of *projected dynamic systems* (PDS) is used to derive a dynamic model for the network, whose stationary points will be the network equilibrium points [221]. Such a model allows us to analyze the network behavior before reaching an equilibrium or in transitions between different equilibrium points.

Theorem (Equilibrium Points): Assume that $\mathscr{K}$ is a convex polyhedron. Then, the equilibrium points of the PDS($\mathscr{K}, F$) coincide with the solutions of VI($\mathscr{K}, F$) [221].

Regarding the affine mapping $\mathbf{F}(\mathbf{p}) = \sigma + \mathbf{Mp}$, the *stationary points* of the following PDS($\mathscr{K}, \sigma + \mathbf{Mp}$), namely,

$$\dot{\mathbf{p}} = \Pi_{\mathscr{K}}(\mathbf{p}, -\mathbf{F}(\mathbf{p})) \tag{4.8}$$

$$\mathbf{p}(t_0) = \mathbf{p}_0 \in \mathscr{K} \tag{4.9}$$

coincide with solutions of the VI($\mathscr{K}, \sigma + \mathbf{Mp}$) problem of (4.7). This PDS provides a dynamic model for the competitive system whose equilibrium behavior is described by the VI. In accordance with the projection operator, $\Pi_{\mathscr{K}}(\mathbf{p}, -\mathbf{F}(\mathbf{p}))$, a point in the interior of $\mathscr{K}$ is projected onto itself, and a point outside of $\mathscr{K}$ is projected onto the closest point on the boundary of $\mathscr{K}$.

To be more precise, we recall the definition of the set of inward normals at $\mathbf{p} \in \mathscr{K}$ as [221]:

$$S(\mathbf{p}) = \{\gamma : \ \|\gamma\| = 1, \langle \gamma, \mathbf{p} - \mathbf{y} \rangle \leq 0, \ \forall \mathbf{y} \in \mathscr{K}\}. \tag{4.10}$$

When $\mathbf{p}(t)$ is in the interior of the feasible set, the projection operator in (4.8) is

$$\Pi_{\mathscr{K}}(\mathbf{p}, -\mathbf{F}(\mathbf{p})) = -\mathbf{F}(\mathbf{p}). \tag{4.11}$$

If $\mathbf{p}(t)$ reaches the boundary of the feasible set, we then have

$$\Pi_{\mathscr{K}}(\mathbf{p}, -\mathbf{F}(\mathbf{p})) = -\mathbf{F}(\mathbf{p}) + z(\mathbf{p})\mathbf{s}^*(\mathbf{p}), \tag{4.12}$$

where

$$\mathbf{s}^*(\mathbf{p}) = \underset{\mathbf{s} \in S(\mathbf{p})}{\operatorname{argmax}} \langle -\mathbf{F}(\mathbf{p}), -\mathbf{s} \rangle \tag{4.13}$$

and

$$z(\mathbf{p}) = \max(0, \langle -\mathbf{F}(\mathbf{p}), -\mathbf{s}^*(\mathbf{p}) \rangle). \tag{4.14}$$

From (4.11) and (4.12), we clearly see that

$$\|\Pi_{\mathscr{K}}(\mathbf{p}, -\mathbf{F}(\mathbf{p}))\| \leq \| - \mathbf{F}(\mathbf{p})\|. \tag{4.15}$$

Figure 4.6 illustrates the function of the projection operator for a two-dimensional case.

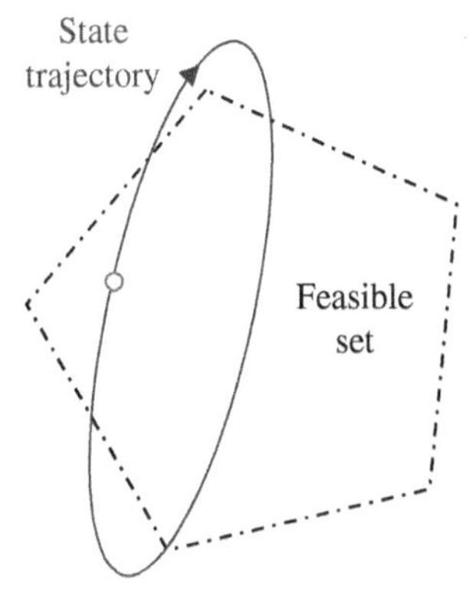

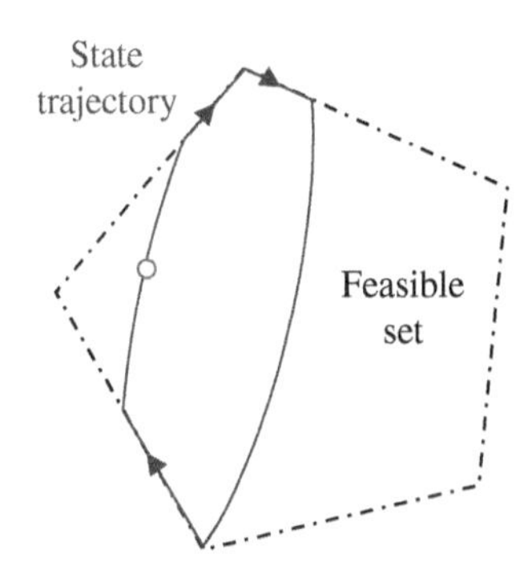

Classic dynamic system Projected dynamic system

FIGURE 4.6 Geometric interpretation of projected dynamic systems. In a projected dynamic system, the state trajectory is confined to the feasible set by a projection operator. The state equation of such systems has a discontinuous right-hand side due to the projection operator. Source: Haykin and Setoodeh (2015) [37]. Reproduced with the permission of IEEE.

The underlying theory of PDS allows the equilibrium problem to be studied in a dynamic model, which enables us not only to study the transient behavior of the network but also to predict it. The associated dynamic model to the equilibrium problem will be realistic if, and only if, there is a unique path from a given initial point, which prompts us to introduce the following:

Theorem (Trajectory Uniqueness): If $\mathbf{F}$ is the initial value problem

$$\dot{\mathbf{p}} = \Pi_{\mathcal{K}}(\mathbf{p}, -\mathbf{F}(\mathbf{p})) \tag{4.16}$$

with

$$\mathbf{p}(t_0) = \mathbf{p}_0 \in \mathcal{K} \tag{4.17}$$

is *Lipschitz* continuous, then for any $\mathbf{p}_0 \in K$, there exists a unique solution $\mathbf{p}(t)$ to the above initial value problem.

Therefore, the *Lipschitz* continuity of the affine mapping $\mathbf{F}(\mathbf{p}) = (\boldsymbol{\sigma} + \mathbf{M}\mathbf{p})$, guarantees the uniqueness of the solution of the initial value problem (4.8) and (4.9) [221]. For the affine mapping $\mathbf{F}(\mathbf{p}) = (\boldsymbol{\sigma} + \mathbf{M}\mathbf{p})$, we have

$$||\mathbf{F}(\mathbf{x}) - \mathbf{F}(\mathbf{y})|| = ||\mathbf{M}(\mathbf{x} - \mathbf{y})||. \tag{4.18}$$

According to the multiplicative property of a matrix norm [222], we can write the following inequality for the right-hand side of (4.18):

$$||\mathbf{M}(\mathbf{x} - \mathbf{y})|| \leq ||\mathbf{M}||.||\mathbf{x} - \mathbf{y}||. \tag{4.19}$$

Moreover, for the Euclidean norm, we have [222]

$$\overline{\sigma}(\mathbf{M}) \leq \|\mathbf{M}\| \leq \sqrt{mn}\,\overline{\sigma}(\mathbf{M}), \tag{4.20}$$

where $\overline{\sigma}(\mathbf{M})$ is the maximum singular value of $\mathbf{M}$. Based on (4.18)–(4.20), we may therefore write

$$\|\mathbf{F}(\mathbf{x}) - \mathbf{F}(\mathbf{y})\| \leq \sqrt{mn}\,\overline{\sigma}(\mathbf{M})\|\mathbf{x} - \mathbf{y}\|. \tag{4.21}$$

The interference channel is a *multiple-input-multiple-output* (MIMO) dynamic system with the state-transition matrix $\mathbf{M}$, in which the transmitted signal by each transmitter on each subcarrier is an input and the received signal by each receiver on each subcarrier is an output. In a MIMO system, the largest gain (amplification) for any input direction is equal to the maximum singular value of the state-transition matrix [222]. The communication channel attenuates the transmitted signals in all directions and, therefore, the Lipschitz continuity is indeed a valid assumption. Hence, there is a unique solution path from a given initial point, and associating a dynamic model to the equilibrium problem is meaningful.

Monotonicity property of the corresponding map provides insight on the characteristics of the equilibrium solution as well as the network stability. Formal definitions of monotonicity and monotone attractors are recalled from [221] and [94], which prompts us to introduce the following definition that embodies four different classes of monotonicity.

Definition (Monotonicity): A mapping $\mathbf{F} : \mathcal{K} \subseteq \mathbb{R}^n \rightarrow \mathbb{R}^n$ is said to be

(a) *monotone* on $\mathcal{K}$ if

$$(\mathbf{F}(\mathbf{x}) - \mathbf{F}(\mathbf{y}))^T(\mathbf{x} - \mathbf{y}) \geq 0,\ \forall \mathbf{x}, \mathbf{y} \in \mathcal{K}; \tag{4.22}$$

(b) *strictly monotone* on $\mathcal{K}$ if

$$(\mathbf{F}(\mathbf{x}) - \mathbf{F}(\mathbf{y}))^T(\mathbf{x} - \mathbf{y}) > 0,\ \forall \mathbf{x}, \mathbf{y} \in \mathcal{K}, \mathbf{x} \neq \mathbf{y}; \tag{4.23}$$

(c) ξ*-monotone* on $\mathcal{K}$ for some $\xi > 1$ if there exists a constant $c > 0$ such that

$$(\mathbf{F}(\mathbf{x}) - \mathbf{F}(\mathbf{y}))^T(\mathbf{x} - \mathbf{y}) \geq c\|\mathbf{x} - \mathbf{y}\|^{\xi},\ \forall \mathbf{x}, \mathbf{y} \in \mathcal{K}; \tag{4.24}$$

(d) *strongly monotone* on $\mathcal{K}$ if there exists a constant $c > 0$ such that

$$(\mathbf{F}(\mathbf{x}) - \mathbf{F}(\mathbf{y}))^T(\mathbf{x} - \mathbf{y}) \geq c\|\mathbf{x} - \mathbf{y}\|^2,\ \forall \mathbf{x}, \mathbf{y} \in \mathcal{K}. \tag{4.25}$$

That is, if $\mathbf{F}$ is 2-monotone on $\mathcal{K}$.

Strong monotonicity implies strict monotonicity, and strict monotonicity implies monotonicity, but the reverse is not true [94].

Definition (Monotone Attractor): Let $\mathcal{K}$ be a closed, convex subset of a Hilbert space [223], where we have the following pair of points:

(a) A point $\mathbf{x}^* \in \mathscr{K}$ is a monotone attractor for the system, if there exists a neighborhood V of $\mathbf{x}^*$ such that the distance $\mathbf{d}(t) = \|\mathbf{x}(t) - \mathbf{x}^*(t)\|$ is a nonincreasing function of t, for any solution $\mathbf{x}(t)$ starting in the neighborhood V.

(b) A point $\mathbf{x}^* \in \mathscr{K}$ is a strict monotone attractor, if the distance $\mathbf{d}(t)$ is decreasing.

The following theorem states the condition under which there exists a unique equilibrium solution for the VI problem [94].

Theorem (Existence and Uniqueness of VI Solution): Let $\mathscr{K} \subseteq \mathbb{R}^n$ be closed convex and $\mathbf{F} : \mathscr{K} \subseteq \mathbb{R}^n \to \mathbb{R}^n$ be continuous.

(a) If $\mathbf{F}$ is strictly monotone on $\mathscr{K}$, then VI($\mathscr{K}, \mathbf{F}$) has at most one solution.

(b) If $\mathbf{F}$ is ξ-monotone on $\mathscr{K}$ for some $\xi > 1$, then VI($\mathscr{K}, \mathbf{F}$) has a unique solution.

Therefore, the VI($\mathscr{K}, \boldsymbol{\sigma} + \mathbf{Mp}$) has at most one solution if $\boldsymbol{\sigma} + \mathbf{Mp}$ is strictly monotone and it has a unique solution, $\mathbf{p}^*$, if $\boldsymbol{\sigma} + \mathbf{Mp}$ is ξ-monotone for some $\xi > 1$ [66]. The local uniqueness of $\mathbf{p}^*$ is not sufficient to guarantee the solvability of the perturbed AVI, but it is important for sensitivity analysis because every unique solution of a VI problem is an *attractor* of all solutions of nearby VIs [94]. Alternatively, we may investigate this issue based on the corresponding PDS. Hence, the following theorem is recalled from [221] about stability of the corresponding PDS.

Theorem (PDS Equilibrium Characteristics): Suppose that $\mathbf{x}^*$ solves VI($\mathscr{K}, \mathbf{F}$).

(a) If the mapping $\mathbf{F}$ is strictly monotone at $\mathbf{x}^*$, then $\mathbf{x}^*$ is a strict monotone attractor for the PDS($\mathscr{K}, \mathbf{F}$).

(b) If the mapping $\mathbf{F}$ is ξ-monotone at $\mathbf{x}^*$ with $\xi < 2$, then $\mathbf{x}^*$ is a finite-time attractor.

(c) If the mapping $\mathbf{F}$ is strongly monotone at $\mathbf{x}^*$, then $\mathbf{x}^*$ is exponentially stable.

Monotonicity of the affine map $\mathbf{Mp} + \boldsymbol{\sigma}$, where $\mathbf{M}$ is not necessarily symmetric, is equivalent to the condition that all of the eigenvalues of $\mathbf{M}$ have nonnegative real parts. Also, strict monotonicity, ξ-monotonicity, and strong monotonicity of $\mathbf{Mp} + \boldsymbol{\sigma}$ as well as the condition that all of the eigenvalues of $\mathbf{M}$ have positive real parts, are all equivalent [94]. The latter condition is equivalent to the statement that $-\mathbf{M}$ is a *Hurwitz* matrix [222]. Therefore, if the matrix $-\mathbf{M}$ is Hurwitz, then exponential stability of the equilibrium solution is guaranteed. It will be clear later that the Hurwitz property of the matrix $-\mathbf{M}$ is also needed to guarantee robust exponential stability of the system in the presence of multiple-time-varying delays [66].

The Hurwitz condition of matrix $-\mathbf{M}$ naturally depends on the topology of the network. Simply stated, if each user's receiver has the proper distance from its own transmitter, which is short compared to its distance from other active transmitters in the network, then it can be guaranteed that the network will reach a stable unique equilibrium [66]. This condition is practically achievable through dynamic spectrum management and spectrum-aware routing as shown in Figures 3.6 and 3.7.

While the dynamic-spectrum manager makes sure that the neighboring transmitters will not use the same set of channels [200], opportunistic-spectrum *ad hoc* routing can guarantee that the distance between receivers and their corresponding transmitters is short enough, compared to their distances from other active transmitters in the network.

In multihop transmission, one or more intermediate nodes along the route between the source and destination nodes forward or relay the transmitted packet to assist the transmission. In this way, the message can be delivered to a destination node, whose distance from the source node is beyond the transmitting range of the source node. Since the distance between source and destination is broken down to several shorter links and the message is transmitted over those short links, multihop transmission can improve energy efficiency in the network. In *ad hoc* networks, where there is no central authority to enforce cooperation between nodes, the nodes may find packet forwarding in conflict with their self-interests. They may refuse to consume their limited power budget to forward other nodes' messages. In such a case, messages should be retransmitted or rerouted through other nodes that are willing to cooperate. Retransmission and rerouting lead to inefficient use of network resources. Two approaches have been proposed to encourage cooperation in wireless ad hoc networks: development of a virtual currency and reputation assignment to each node. These approaches may have the potential for fraud and misinterpretation, respectively. Also, both of them increase the network traffic overhead and computational complexity at each node.

Figure 4.7 shows a two-player packet-forwarding game [80, 224]. Each source tries to send packets to its corresponding destination and relies on the other source to forward its packet. If sources do not cooperate, destinations will not receive the transmitted packets and the power that each source consumed for packet transmission will be wasted. The payoff matrix is illustrated in Figure 4.8. The packet-forwarding game is basically the so-called *prisoner's dilemma*, which is played repeatedly. The relaying game, which is the packet-forwarding game with taking account of the physical channel's condition, may also be considered in the way shown in Figure 4.9 [224]. It is assumed that direct transmission between each source and its corresponding destination is possible and payoffs are based on the amount of energy that could be saved using multihop transmission instead of direct transmission.

Using toy problems, the authors in [224] tried to find out if cooperation can emerge in packet-forwarding and relaying games without extrinsic mechanisms. They concluded that if in a repeated game with uncertain ending, credible punishment exists and players are sufficiently patient in the sense that they appreciate long-term payoffs, then natural cooperation can emerge.

4.2.1.1 Computer Experiment: PDS Using the test bed described in the previous chapter, simulation results are now presented to support theoretical underpinnings of this section. Network dynamics are simulated and the solution stability is studied under system perturbation. Numerical values for parameters are chosen in the same way that was described previously.

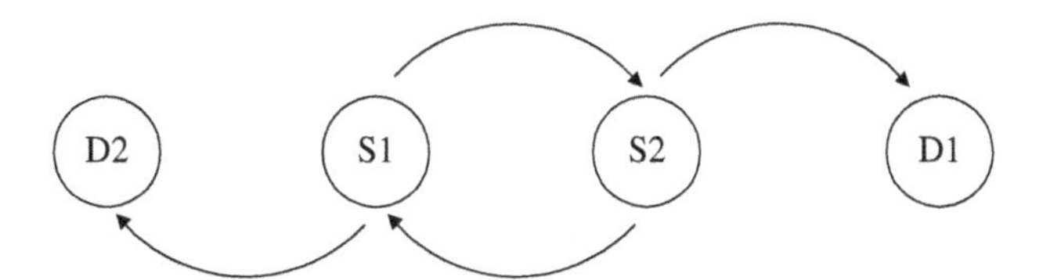

FIGURE 4.7 A two-player packet-forwarding scenario.

	S2 does not forward	S2 forwards
S1 does not forward	(0,0)	(1,–c)
S1 forwards	(–c,1)	(1–c,1–c)

FIGURE 4.8 Payoff matrix for a symmetric two-player packet-forwarding game.

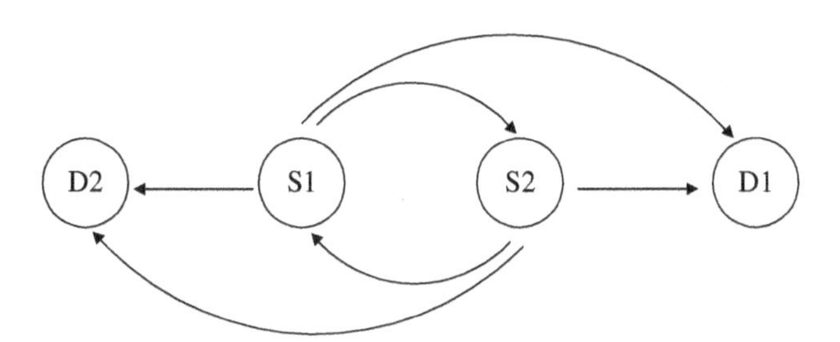

FIGURE 4.9 A two-player relaying scenario.

To study the transient behavior of a cognitive radio network, a scenario is considered for three users and three subcarriers, so that we may arrive at insightful conclusions. Moreover, three are chosen merely for the sake of illustration. It is assumed that all the users update their power vectors simultaneously under the assumption that the network experiences the worst-case interference conditions.

Figure 4.10 depicts state trajectories for three users obtained from a discrete-time approximation of the PDS by solving the quadratic programming described in (4.84), when the following sequence of events happens. First, all three subcarriers are idle and can be used by the secondary users. Therefore, the state trajectories evolve in the three-dimensional space (i.e., $p_1^i p_2^i p_3^i$ space). Then, the second subcarrier is no longer available and state trajectories enter the two-dimensional space and evolve in $p_1^i p_3^i$ plane. Subsequently, the same thing happens to the third subcarrier and state trajectories evolve in one-dimensional space (i.e., p_1^i line). After a while, subcarrier three becomes idle and therefore available again. Thus, the state trajectories enter from p_1^i line to $p_1^i p_3^i$ plane. When subcarrier two becomes available again, state trajectories enter from $p_1^i p_3^i$ plane to $p_1^i p_2^i p_3^i$ space.

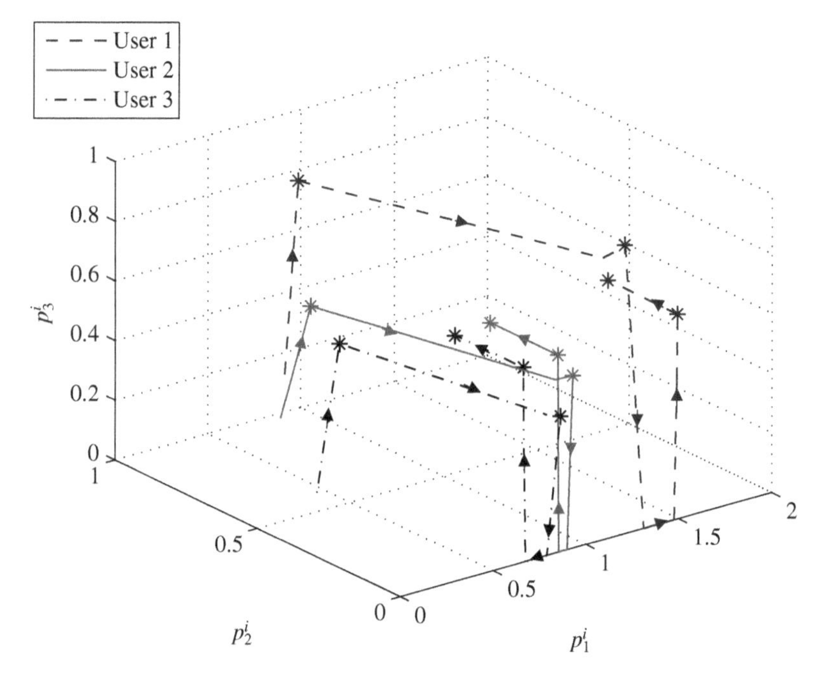

FIGURE 4.10 Power trajectories for a network of three users with three available subcarriers obtained from the associated PDS, when both the interference gains and the number of subcarriers change by time. Direction of evolution of states and the achieved equilibrium points are shown by arrows and asterisks, respectively. Trajectories enter lower dimensional spaces when spectrum holes disappear and then, go back to higher-dimensional spaces again, when new spectrum holes are available. When the second subcarrier is not idle, trajectories enter $p_1^i p_3^i$ plane and when the third subcarrier is not also idle anymore, trajectories enter p_1^i line. After a while when third and then second subcarriers become available again, state trajectories go back to $p_1^i p_3^i$ plane and then $p_1^i p_2^i p_3^i$ space. Setoodeh and Haykin (2009) [66]. Reproduced with the permission of IEEE.

It is obvious that the power trajectories enter from higher to lower dimensional spaces according to the number of available subcarriers, and again they go back to higher dimensional spaces when users have access to more subcarriers, which is what should happen during a successful operation. The achieved equilibrium points for different users as they exist between occurrences of the mentioned events are shown by asterisks on their state trajectories. Also, arrows in Figure 4.10 show the direction of evolution of states for different users.

4.2.1.2 Computer Experiment: Sensitivity Analysis To study the solution stability via simulation, the system is perturbed and the equilibrium point of the perturbed system is calculated. The interference-gain matrix and the noise vector are, respectively, perturbed as $\mathbf{M} + w_{\mathbf{M}}\Delta\mathbf{M}$ and $\boldsymbol{\sigma} + w_{\sigma}\Delta\boldsymbol{\sigma}$, where $w_{\mathbf{M}}$ are w_{σ} are weights. The perturbation terms $\Delta\mathbf{M}$ and $\Delta\boldsymbol{\sigma}$ are chosen in the same way that $\mathbf{M}$ and $\boldsymbol{\sigma}$ were chosen, respectively. Results at three different subcarriers are shown separately in Figure 4.11. As the perturbation terms decay (i.e., the weights $w_{\mathbf{M}}$ are w_{σ} move toward zero) and the perturbed system approaches the original one, behavior of the perturbed system converges to the solution of the original system, which is shown by asterisks in Figure 4.11. The arrows show the direction in which the solution of the perturbed system converges to the solution of the original system. This experiment validates the notion of solution stability that was discussed previously.

4.2.2 Cognitive Radio Network Viewed as a Hybrid Dynamic System

As mentioned previously, the cognitive radio network is a hybrid dynamic system with both continuous and discrete dynamics. Changes occur in the network due to discrete events such as the appearance and disappearance of users and spectrum holes, as well as continuous dynamics described by differential equations that govern the evolution of transmit power vectors of cognitive radio users over time. When the conditions change due to these kinds of discrete events, each user will have to solve a new optimization problem similar to the one described in (3.82); the network deviates from the achieved equilibrium point, and it is desirable to converge to a new one reasonably fast. Also, the occurrence of an event such as a change in the number of users or available subcarriers will change the parameters $\boldsymbol{\sigma}$ and $\mathbf{M}$. Accordingly, the problem is formulated in terms of an ensemble of subsystems and the global state space is addressed as follows [66]:

- partitioned into polyhedral regions follow the varying realizations of the network at different time intervals and
- an affine state equation

  $$\dot{\mathbf{p}} = -\mathbf{F}(\mathbf{p}) = -(\boldsymbol{\sigma} + \mathbf{M}\mathbf{p}), \ \forall \mathbf{p}(t) \in \mathscr{K} \tag{4.26}$$

  is associated with each polyhedral region that governs the evolution of state trajectory in that region.

It follows therefore that the whole network can be modeled as a constrained *piecewise affine* (PWA) system [225]:

$$\dot{\mathbf{p}} = -\mathbf{M}(\boldsymbol{v})\mathbf{p} - \boldsymbol{\sigma}(\boldsymbol{v}), \ \forall \mathbf{p}(t) \in \mathscr{K}(\boldsymbol{v}), \tag{4.27}$$

where $\boldsymbol{v}$ is a key vector that is a function of time and discrete events; moreover, it describes which affine subsystem is currently a valid representation of the network [226].

The stationary points of each one of these dynamic subsystems coincide with the equilibrium points of the corresponding game resulting from solving the related optimization problems. In summary, the occurrence of discrete events changes the equilibrium point and causes the state trajectory to deviate from an equilibrium point and, therefore, converge to another equilibrium point. Each one of these equilibrium points may have a *region of attraction* around it, such that if the system is perturbed, then the solution remains in that region close enough to the solution of the unperturbed system. This issue will be studied in the following subsection.

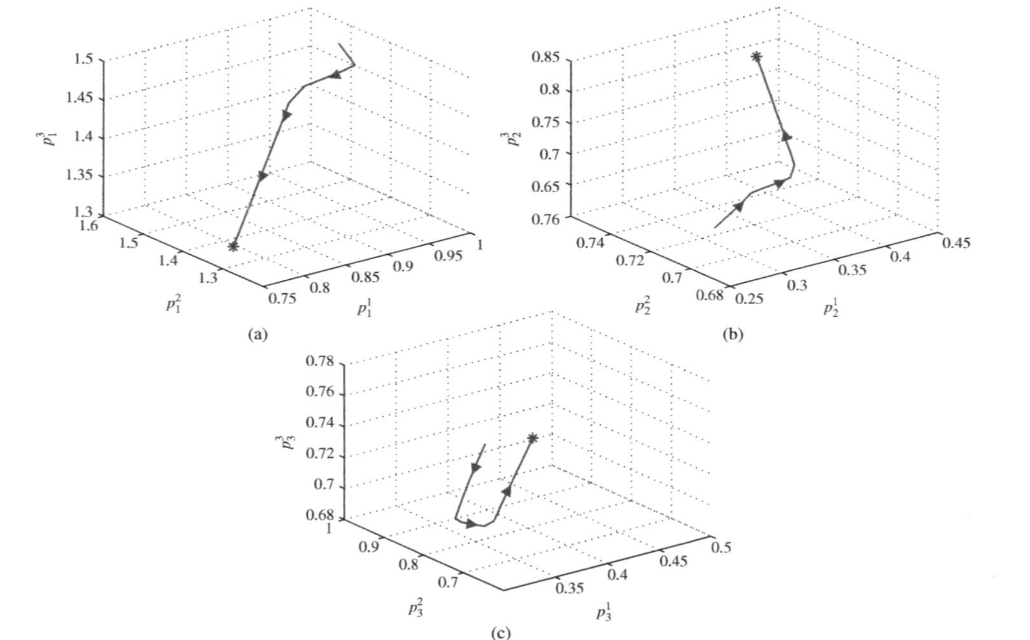

FIGURE 4.11 Solution stability analysis. Solution of the perturbed system converges to the solution of the original system (shown by asterisks) as the perturbed system approaches the original system. Results are depicted for different subcarriers separately: (a) subcarrier 1, (b) subcarrier 2, and (c) subcarrier 3. Arrows show the direction of convergence. Setoodeh and Haykin (2009) [66]. Reproduced with the permission of IEEE.

4.2.3 Network Stability in the Presence of Uncertainty and Time Delay

To reiterate, the perception–action cycle is a distinctive characteristic of cognitive radio. In order to establish such a cycle between any pair of cognitive radio users in the network, denoted by A and B, there would have to be the following two channels:

- Feedback channel, which links the transmitter of user B to the receiver of user A.
- Communication (forward channel), which links the transmitter of user A to the receiver of user B.

In such a scenario, user A is speaking with user B. Naturally, for a two-way communication, each user is equipped with a transceiver. There are various ways, by means of which the feedback channel can be established. Perhaps, the simplest way of doing it is to use unlicensed bands [46]. Also, in order to be conservative in consuming the precious bandwidth that can be used for data transmission, the feedback channel should be low rate and quantized.

Feedback naturally introduces delay in the control loop and different transmitters may receive statistics of noise and interference with different time delays. Moreover, the sporadic feedback causes the users to adopt outdated statistics to update their power vectors. The time-varying delay in the control loop degrades the performance and may cause stability problems. Analysis of stability and control of time-delay systems is a topic of practical interest. Robust stability of the system under time-varying delays is the focus of this subsection [66].

The dynamic model of the previous subsections can be used to find out if the network is able to achieve a *retarded equilibrium*, which is stable. If an equilibrium point is not stable, the system may not be able to maintain that state long enough because of perturbations, and there is the potential possibility that an equilibrium cannot be established.

The dynamic model of (4.8) will be a projected dynamic system with delay (PDSD) [227] in the form of the following *functional differential equation* (FDE) [228–230]:

$$\dot{\mathbf{p}}(t) = \Pi_{\mathscr{K}}(\mathbf{p}(t), -\mathbf{F}_d(\mathbf{p})), \tag{4.28}$$

where $\mathbf{F}_d$ may be written as

$$\mathbf{F}_d(\mathbf{p}) = \begin{bmatrix} \mathbf{F}^1(\mathbf{p}^1(t), \mathbf{p}^{-1}(\mathscr{S}_t)) \\ \vdots \\ \mathbf{F}^i(\mathbf{p}^i(t), \mathbf{p}^{-i}(\mathscr{S}_t)) \\ \vdots \\ \mathbf{F}^m(\mathbf{p}^m(t), \mathbf{p}^{-m}(\mathscr{S}_t)) \end{bmatrix} \tag{4.29}$$

and where $\mathbf{p}^{-i}(\mathscr{S}_t)$ denotes a continuous-time asynchronous adjustment scheme similar to (3.86).

Let the given initial time be t_0. In order to determine the continuous solution, $\mathbf{p}(t)$ of (4.28) for $t > t_0$, we need to know a continuous *initial function*, $\boldsymbol{\phi}(t)$, where $\mathbf{p}(t) = \boldsymbol{\phi}(t)$ for $t_0 - \tau^{i,j} \le t \le t_0$, $\forall i, j = 1, \ldots, n$. The initial function may be obtained from measurements. Moreover, since the system described in (4.28) and (4.29) is a multiple-delay system, each deviation defines an initial set $\Psi_{t_0}^{i,j}$, consisting of the point t_0 and those values $t - \tau^{i,j}(t)$ for which $t - \tau^{i,j}(t) < t_0$ when $t \ge t_0$ [231].

Therefore, the initial condition for the system described in equation (4.28) is

$$\mathbf{p}(\theta) = \boldsymbol{\phi}(\theta), \quad \forall \theta \in \Psi_{t_0}, \tag{4.30}$$

where $\boldsymbol{\phi} : \Psi_{t_0} \mapsto \mathbb{R}^{m \times n}$ is a continuous norm-bounded initial function [231, 232] and

$$\begin{aligned} \Psi_{t_0} &= \bigcup_{i,j=1, i \ne j}^{m} \Psi_{t_0}^{i,j} \\ &= \bigcup_{i,j=1, i \ne j}^{m} \{t \in \mathbb{R} : t = \kappa - \tau^{i,j}(\kappa) \le 0, \kappa \ge t_0\}. \end{aligned} \tag{4.31}$$

The $\mathbf{F}_d(\mathbf{p})$ in (4.29) can be written as the following summation:

$$\begin{aligned} \mathbf{F}_d(\mathbf{p}) = \mathbf{p}(t) &+ \sum_{i=1}^{m} \sum_{j=1, \ne i}^{m} \mathbf{M}_d^{ij} \mathbf{p}(t - \tau^{i,j}(t)) \\ &+ \sum_{i=1}^{m} \sum_{j=1, \ne i}^{m} \Delta \mathbf{M}_d^{ij} \mathbf{p}(t - \tau^{i,j}(t)) + \rho(t), \end{aligned} \tag{4.32}$$

where $\mathbf{M}_d^{ij}$ is obtained by replacing all the blocks in $\mathbf{M}$ except $\mathbf{M}^{ij}$ by $n \times n$ zero matrices, and $\Delta \mathbf{M}_d^{ij}$ is a perturbation in $\mathbf{M}_d^{ij}$. The term ρ is the combined effect of the background noise in both the forward and feedback channels.

Therefore, the associated constrained affine system that governs the network's dynamics is described by the differential equation:

$$\begin{aligned} \dot{\mathbf{p}}(t) = -\mathbf{p}(t) &- \sum_{i=1}^{m} \sum_{j=1, \ne i}^{m} \mathbf{M}_d^{ij} \mathbf{p}(t - \tau^{i,j}(t)) \\ &- \sum_{i=1}^{m} \sum_{j=1, \ne i}^{m} \Delta \mathbf{M}_d^{ij} \mathbf{p}(t - \tau^{i,j}(t)) - \rho(t). \end{aligned} \tag{4.33}$$

$\forall \mathbf{p}(t) \in K$, which is a multiple-time-varying-delay system with uncertainty. It can be written as

$$\begin{aligned} \dot{\mathbf{p}}(t) = -\mathbf{p}(t) &- \sum_{\ell=1}^{m(m-1)} \mathbf{M}_d^{\ell} \mathbf{p}(t - \tau^{\ell}(t)) \\ &- \sum_{\ell=1}^{m(m-1)} \Delta \mathbf{M}_d^{\ell} \mathbf{p}(t - \tau^{\ell}(t)) - \rho(t). \end{aligned} \tag{4.34}$$

This reformulation is an instance of the general systems that were studied in [232]. Following the approach of [232], we assume that $\forall t \geq t_0$, the time-varying delays $\tau^{\ell}(t)$ satisfy

$$\tau^{\ell}(t) \leq \tau(t) \leq \overline{\tau} \tag{4.35}$$

$$\dot{\tau} \leq \delta < 1, \tag{4.36}$$

where $\overline{\tau} > 0$ and $\delta \geq 0$. $\tau(t)$ is strictly positive continuous differentiable function. Also, the uncertainties are assumed to be bounded for all $\mathbf{p}$ and at all times, such that the following pair of conditions hold:

$$||\rho(t)|| \leq b_d ||\mathbf{p}(t)|| \tag{4.37}$$

and

$$||\Delta \mathbf{M}_d^{\ell}(t)\mathbf{p}(t))|| \leq b_d^{\ell} ||\mathbf{p}(t)||, \tag{4.38}$$

where $b_d \geq 0$ and $b_d^{\ell} \geq 0$. If there exist $\zeta \geq 1$ and $\lambda > 0$ such that

$$||\mathbf{p}(t)|| \leq \zeta \sup_{\theta \in \Psi_{t_0}} \{||\mathbf{p}(\theta)||\} e^{-\lambda(t-t_0)}, \tag{4.39}$$

then the uncertain time-delay system of (4.34) is said to be robustly exponentially stable with a decay rate of λ. In other words, the trivial solution, $\mathbf{p} = 0$, of the system, is exponentially stable with a decay rate of λ for all admissible uncertainties [232].

Recognizing that

$$\mathbf{I} + \sum_{i=1}^{m(m-1)} \mathbf{M}_d^{\ell} = \mathbf{M}, \tag{4.40}$$

we may conclude the robust exponential stability of the network from Theorem 4 of [232], which is repeated here with some modification to conform to our problem.

Theorem (Robust Exponential Stability): Consider the system (4.34) with initial condition (4.30), and assume that $-\mathbf{M}$ is a Hurwitz stable matrix satisfying

$$||e^{\mathbf{M}t}|| \leq ce^{-\eta t} \tag{4.41}$$

for some real numbers $c \geq 1$ and $\eta > 0$. In the left-hand side of the above equation, e denotes a "matrix" exponential operator. If the inequality

$$\frac{c}{\eta}\left[\overline{\tau} \sum_{\ell=1}^{m(m-1)} (\mu_1^{\ell} + \mu_2^{\ell}) + b_d + \sum_{\ell=1}^{m(m-1)} b_d^{\ell}\right] < 1 \tag{4.42}$$

holds, then the transient response of $\mathbf{p}(t)$ satisfies

$$||\mathbf{p}(t)|| \leq \zeta \sup_{\theta \in \Psi_{t_0}} \{||\boldsymbol{\phi}(\theta)||\} e^{-\rho \int_{t_0}^{t} \frac{d\theta}{\tau(\theta)}}, \ \forall t \geq t_0, \zeta \geq 1, \tag{4.43}$$

where

$$\mu_1^\ell = \|\mathbf{M}_d^\ell\| + \|\mathbf{M}_d^\ell\| b_d \tag{4.44}$$

$$\mu_2^\ell = \sum_{j=1}^{m(m-1)} \|\mathbf{M}_d^\ell \mathbf{M}_d^j\| + \|\mathbf{M}_d^\ell\| \sum_{j=1}^{m(m-1)} b_d^j \tag{4.45}$$

and $\rho > 0$ is the unique positive solution of the transcendental equation

$$1 - \frac{c}{\eta} b_d - \frac{\rho}{\eta \tau(0)} = \mu_3 \frac{c}{\eta} e^{\frac{\rho}{1-\delta}}, \tag{4.46}$$

where

$$\mu_3 = \overline{\tau} \sum_{\ell=1}^{m(m-1)} \mu_1^\ell + \overline{\tau} e^{\frac{\rho}{1-\delta}} \sum_{\ell=1}^{m(m-1)} \mu_2^\ell + \sum_{\ell=1}^{m(m-1)} b_d^\ell. \tag{4.47}$$

Furthermore, the system described by (4.34) and (4.30) is robustly exponentially stable with a decay rate $\rho/\overline{\tau}$.

The left-hand side of the transcendental equation (4.46) is a continuous decreasing function of ρ and its right-hand side is a continuous increasing function, and by virtue of (4.42) at $\rho = 0$, the right-hand side is less than the left-hand side. Therefore, (4.46) has a unique positive solution as desired.

4.2.3.1 Computer Experiment: PDSD Simulation results are now presented for the network dynamics in the presence of multiple-time-varying delays. When delays introduced by the feedback channels are considered, it may take longer for both the original system and the perturbed systems to achieve an equilibrium. Under the conditions mentioned in this section, the robust exponential stability of the system is guaranteed and similar results are obtained in simulations for the time-delay cases with constraints on delays.

Simulation results for the above network of three users and three potentially available subcarriers, with a similar sequence of events, mentioned in Subsection 4.2.1, are repeated with asynchronous adjustment scheme. In the beginning, all three subcarriers are idle and can be used by secondary users. Then, the second subcarrier is no longer available, and after that the same thing happens to the third subcarrier. After a while, subcarriers two and then three become idle and therefore available again. Power trajectories and achieved equilibrium points are shown in Figure 4.12. Furthermore, Figure 4.13 depicts the random delays in adjustment schemes, $\tau^i(t)$, used by different users, which shows that most of the time users have used outdated information to update their power vectors. The results confirm ability of the system to achieve retarded equilibria under the conditions given in *Theorem* (*Robust Exponential Stability*). By increasing the delay, the performance of the system will degrade and eventually the system becomes unstable.

In summary, simulations were conducted to demonstrate the concept of solution stability. The system was perturbed and its equilibrium solution was calculated. By

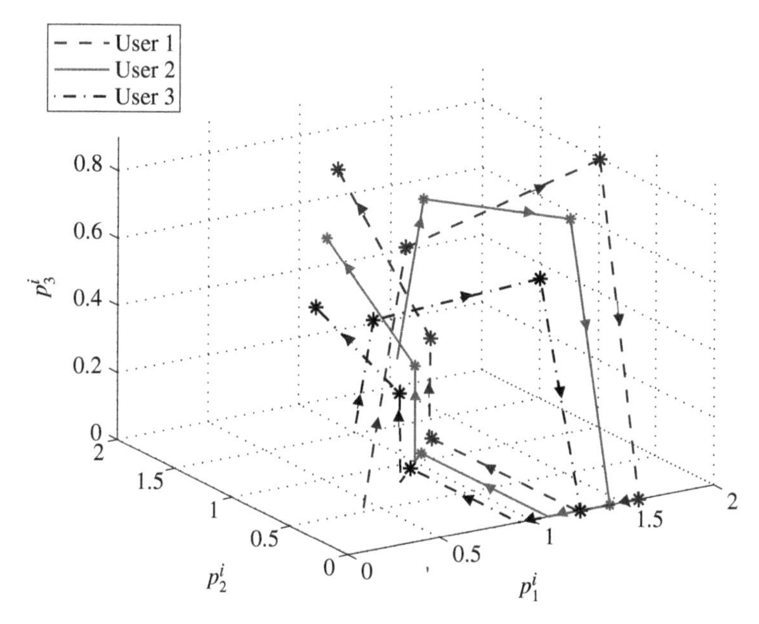

FIGURE 4.12 Power trajectories for a network of three users with three available subcarriers obtained from the associated multiple-time-varying-delay PDS with uncertainty, when both the interference gains and the number of subcarriers change by time. Direction of evolution of states and the achieved equilibrium points are shown by arrows and asterisks, respectively. Trajectories enter lower dimensional spaces when spectrum holes disappear and then, go back to higher-dimensional spaces again, when new spectrum holes are available. When the second subcarrier is not idle, trajectories enter $p_1^i p_3^i$ plane and when the third subcarrier is not also idle anymore, trajectories enter p_1^i line. After a while when second and then third subcarriers become available again, state trajectories go back to $p_1^i p_2^i$ plane and then $p_1^i p_2^i p_3^i$ space. Setoodeh and Haykin (2009) [66]. Reproduced with the permission of IEEE.

decaying the perturbation terms, the equilibrium solution of the perturbed system converged to the equilibrium solution of the original system. The ability of the dynamic model, obtained using the PDS theory, was validated by simulations for both delay-free and multiple-time-varying-delay cases. The results presented here show that by appearance and disappearance of spectrum holes, the state trajectory of the network enters higher and lower dimensional subspaces in the global state space, respectively.

4.2.4 Double-layer Dynamics of Cognitive Radio Networks

A cognitive radio network may be characterized by different timescales perhaps with different orders of magnitude. Since cognitive radio is aimed at spatiotemporal reuse of spectrum, communication patterns of primary users dictate how cognitive radio users should behave. Hence, communication patterns of cognitive radio users must have faster dynamics compared to those of primary users. This can also be interpreted as a hierarchy of time constants. Slow and fast transients in the network are,

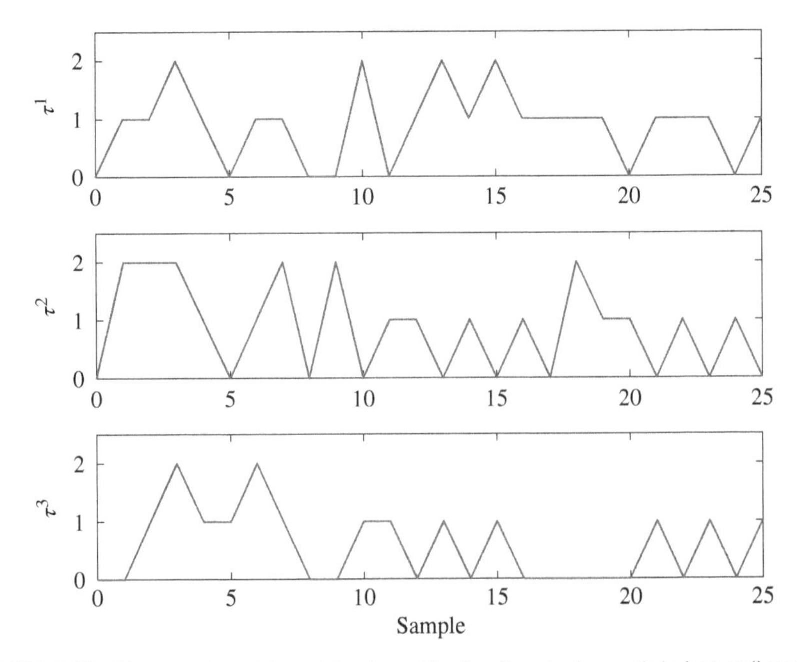

FIGURE 4.13 Time-varying delays introduced by feedback channels in transmit power control loops for a network of three users. Setoodeh and Haykin (2009) [66]. Reproduced with the permission of IEEE.

respectively, associated with large and small time constants on the time axis. In order to study a set of events that take place in a network, we need to use a window of observation to focus on a limited range of timescales, which are associated with the events of interest. In a cognitive radio network, the window of observation distinguishes between three timescales [233]:

- Slow relaxation times: Their corresponding processes can be considered stationary over the observation times.
- The timescale that is of interest.
- Fast relaxation times: It is assumed that their corresponding processes can be eliminated from the dynamic description of the network because they settle fast.

In this way, the network dynamics can be decomposed in time. For a thorough characterization of network dynamics, all time constants must be studied by sliding the window of observation over the time axis.

Regarding the above three timescale categories, a three-dimensional space is presented in Figure 4.14. A change in communication patterns of primary users corresponds to a rapid motion from a two-dimensional space to another one. In each one of the two-dimensional spaces, dynamic spectrum management and transmit power control are associated with the two axes of Figure 4.14. Dynamic spectrum manager

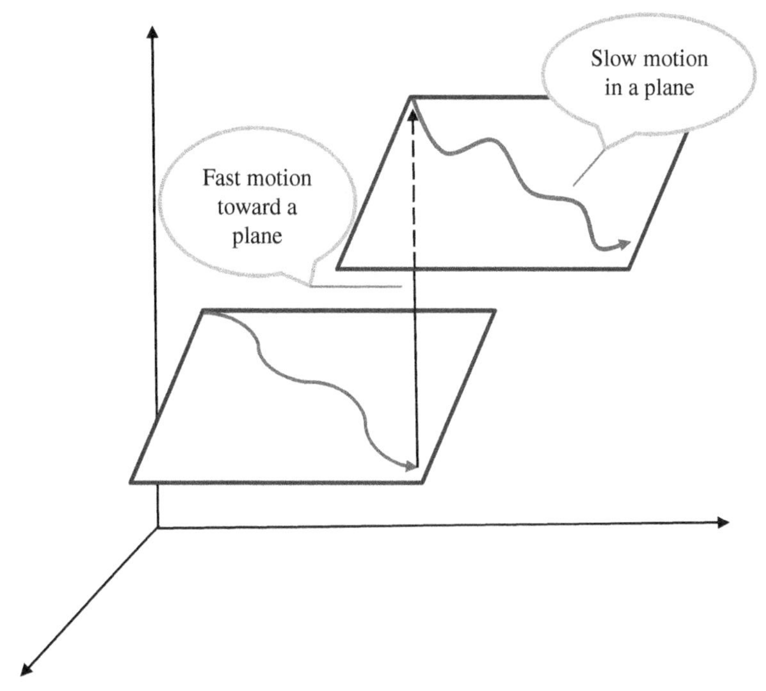

FIGURE 4.14 Timescale decomposition in a three-dimensional space. A change in communication patterns of primary users is viewed as a discrete event and corresponds to a rapid motion from a two-dimensional space to another one. In each one of the two-dimensional spaces, dynamic spectrum management and transmit power control are associated with the two axes. Source: Haykin and Setoodeh (2015) [37]. Reproduced with the permission of IEEE.

selects a set of suitable channels for communication and transmit power controller adjusts the transmit power over the chosen channels to provide an acceptable level of quality of service despite noise and interference [37].

Therefore, a cognitive radio network is indeed a multilayer dynamic system. A model is built that can be used as a testing tool for policy forecast and incorporates time evolution as the life span (control horizon) of a given policy. Two types of time dependency are studied: time-dependent equilibria and time-dependent behavior away from the predicted curve of equilibria. Theories of evolutionary variational inequalities and projected dynamic systems on Hilbert spaces are used to study these two types of time dependency.

Regarding the fact that dynamics play a central role in cognitive radio networks, the joint feasible set of the active users in the network is time varying in nature, and the finite-dimensional feasible set $\mathscr{K}$ captures a snapshot of the underlying dynamics with a time-varying feasible set. Regarding the continuous nature of time, we need to deal with infinite-dimensional feasible sets, if we consider time explicitly in the structure of the feasible set. Hence, the developed model should be extended to *Hilbert spaces*. Hilbert space extends the results of vector algebra and calculus to spaces with any finite or infinite number of dimensions. The real space ℓ_2 is a Hilbert space.

In this section, the previous results are extended to study the equilibrium behavior of the network by explicitly considering time in the AVI-based model. Since the time-varying nature of the network's feasible set is explicitly considered in formulating the model, theory of time-dependent variational inequality or evolutionary VI (EVI) should be employed to obtain an equilibrium model for the network. The EVI-based model gives a *curve of equilibria* over a time interval of interest $[0, T]$. The predictive model will provide a reasonable estimate for T.

By considering time as an additional scalar parameter, the joint feasible set will be the following subset of the Hilbert space $\ell_2([0, T], \mathbb{R}^{m\times n})$:

$$\mathscr{K} = \bigcup_{t\in[0,T]} \mathscr{K}_t, \tag{4.48}$$

where $\mathscr{K}_t$ is the feasible set described previously in this section. The network whose feasible set is described by $\mathscr{K}_t$ at a specific time instant $t \in [0, T]$ is a snapshot of the dynamic network with the time-varying feasible set (4.48) at that particular time instant. The EVI-based equilibrium model of the network may therefore be stated as follows: find the $\mathbf{p}^* \in \mathscr{K}$ such that the following condition

$$\int_0^T (\mathbf{p} - \mathbf{p}^*)^T(\sigma + \mathbf{M}\mathbf{p}^*)\mathrm{d}t \geq 0, \ \forall \mathbf{p} \in \mathscr{K} \tag{4.49}$$

holds [223]. In what follows, the results of [66] are further extended to Hilbert spaces in order to study the equilibrium states of the network obtained from the EVI in a dynamic framework [211].

A generalization of the theory of PDS on Hilbert spaces is used to model the transient behavior of the network, whose equilibrium condition is described by the EVI. Two distinct time frames are considered: large-scale time frame t and small-scale time frame τ. There is a PDS corresponding to each $t \in [0, T]$, which is denoted by PDS_t. However, the evolution time variable for PDS_t is denoted by τ, which is different from time t. PDS_t describes time evolution of the state trajectory of the system toward an equilibrium point on the curve of equilibria corresponding to the moment t. The following PDS [223]

$$\frac{\mathrm{d}\mathbf{p}(.,\tau)}{\mathrm{d}\tau} = \Pi_{\mathscr{K}}(\mathbf{p}(.,\tau), -\mathbf{F}(\mathbf{p}(.,\tau))) \tag{4.50}$$

with the initial condition

$$\mathbf{p}(.,0) = \mathbf{p}_0(.) \in \mathscr{K} \tag{4.51}$$

is established as a dynamic model for the network, which governs the transient behavior of the network preceding the attainment of an equilibrium. The stationary points of the PDS coincide with equilibrium points of the corresponding EVI problem. It follows therefore that the associated dynamic model to the equilibrium problem will be realistic if, and only if, there is a unique solution path from a given initial point, which is guaranteed by the Lipschitz continuity of **F** [223].

Theorem (Trajectory Uniqueness in Hilbert Space): Let H be a Hilbert space and K be a nonempty, closed, convex subset. Let $F : \mathscr{K} \rightarrow H$ be a Lipschitz-continuous vector field and $\mathbf{p}_0 \in \mathscr{K}$. Then, the initial value problem

$$\frac{\mathrm{d}\mathbf{p}(\tau)}{\mathrm{d}\tau} = \Pi_{\mathscr{K}}(\mathbf{p}(\tau), -\mathbf{F}(\mathbf{p}(\tau))), \ \mathbf{p}(0) = \mathbf{p}_0 \in \mathscr{K} \tag{4.52}$$

has a unique absolutely continuous solution on the interval $[0, \infty)$.

Since monotonicity properties of the underlying vector field of EVI/PDS play a key role, we may say that strict monotonicity and Lipschitz continuity of **F** on $\mathscr{K}$ guarantee existence of a unique equilibrium solution [223].

Theorem (Existence and Uniqueness of EVI/PDS$_t$ Equilibria): If $\mathbf{F}(\mathbf{p}) = \sigma + \mathbf{M}\mathbf{p}$ is strictly monotone and Lipschitz continuous on $\mathscr{K}$, then there exists $\mathbf{p}^* \in K$ such that

- $\mathbf{p}^*$ uniquely solves the EVI problem
- $\mathbf{p}^*$ uniquely solves $\Pi_{\mathscr{K}}(\mathbf{p}(.,\tau), -\mathbf{F}(\mathbf{p}(.,\tau))) = 0$

In [66], we discussed the following scenario: If in a cognitive radio network, the distance between receivers and their corresponding transmitters is short enough compared to their distances from other active transmitters in the network, then the strict monotonicity condition is satisfied and, therefore, the EVI will have a unique equilibrium. The corresponding solution of the EVI problem coincides with the stationary point of the corresponding PDS. Now, we seek the answers to the following two questions [211]:

(i) If the initial state of the network is close to an equilibrium (i.e., if the competitive game starts near an equilibrium), will the state trajectory remain in a neighborhood of the equilibrium?

(ii) Starting from an initial state, will the state trajectory asymptotically approach an equilibrium and at what rate?

In EVI, monotonicity establishes the essential conditions for the existence and uniqueness of the solutions. In PDS, monotonicity is used to study stability of the perturbed system. The following definitions are recalled from [223].

Definition (Pseudo-monotonicity): A mapping **F** is called

(a) *pseudo-monotone* on $\mathscr{K}$ if $\forall \mathbf{x}, \mathbf{y} \in \mathscr{K}$

$$\langle \mathbf{F}(\mathbf{x}), \mathbf{y} - \mathbf{x} \rangle \geq 0; \Longrightarrow \langle \mathbf{F}(\mathbf{y}), \mathbf{y} - \mathbf{x} \rangle \geq 0 \tag{4.53}$$

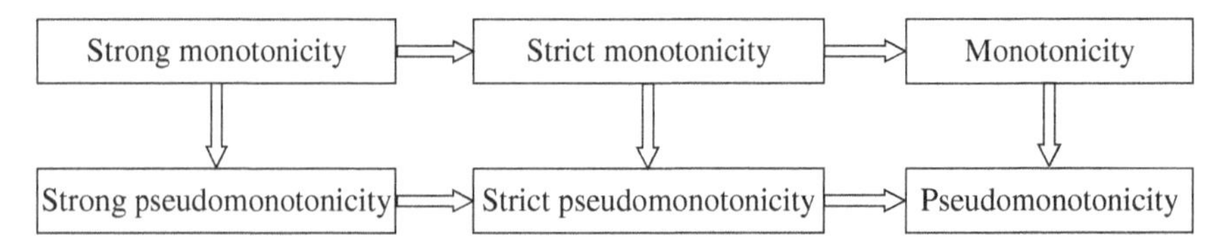

FIGURE 4.15 Relationship between different notions of monotonicity. Strong monotonicity implies strict monotonicity, which, in turn, implies monotonicity. By the same token, strong pseudo-monotonicity implies strict pseudo-monotonicity, which, in turn, implies pseudo-monotonicity. Furthermore, strong monotonicity, strict monotonicity, and monotonicity imply strong pseudo-monotonicity, strict pseudo-monotonicity, and pseudo-monotonicity, respectively. Source: Haykin and Setoodeh (2015) [37]. Reproduced with the permission of IEEE.

(b) *strictly pseudo-monotone* on $\mathcal{K}$ if $\forall \mathbf{x}, \mathbf{y} \in \mathcal{K}, \mathbf{x} \neq \mathbf{y}$

$$\langle \mathbf{F}(\mathbf{x}), \mathbf{y} - \mathbf{x} \rangle \geq 0; \Longrightarrow \langle \mathbf{F}(\mathbf{y}), \mathbf{y} - \mathbf{x} \rangle > 0 \tag{4.54}$$

(c) *strongly pseudo-monotone* on $\mathcal{K}$ if there exists a constant $c > 0$ such that $\forall \mathbf{x}, \mathbf{y} \in \mathcal{K}, \mathbf{x} \neq \mathbf{y}$

$$\langle \mathbf{F}(\mathbf{x}), \mathbf{y} - \mathbf{x} \rangle \geq 0 \Longrightarrow \langle \mathbf{F}(\mathbf{y}), \mathbf{y} - \mathbf{x} \rangle \geq c\|\mathbf{x} - \mathbf{y}\|^2. \tag{4.55}$$

The implications between the different monotonicity notions are shown in Figure 4.15 [234]. Strong monotonicity implies strict monotonicity, which, in turn, implies monotonicity. By the same token, strong pseudo-monotonicity implies strict pseudo-monotonicity, which, in turn, implies pseudo-monotonicity. Furthermore, strong monotonicity, strict monotonicity, and monotonicity imply strong pseudo-monotonicity, strict pseudo-monotonicity, and pseudo-monotonicity, respectively.

Theorem (PDS$_t$ Equilibria Characteristics): Assume $\mathbf{F} : \mathcal{K} \rightarrow \ell_2([0, T], \mathbb{R}^{m \times n})$ is Lipschitz continuous on $\mathcal{K}$

- If $\mathbf{F}$ is strictly pseudo-monotone on $\mathcal{K}$, then the unique curve of equilibria is a strict monotone attractor.
- If $\mathbf{F}$ is strongly pseudo-monotone on $\mathcal{K}$, then the unique curve of equilibria is exponentially stable and an attractor.

To address the two questions pertaining to a disturbed network, we first appeal to properties of the ℓ_2-norm, according to which the network is expected to evolve uniformly toward its equilibrium on the curve of possible equilibria for practically all $t \in [0, T]$. Here again, in [66] and [211], we discussed the fact that the Hurwitz condition of matrix $-\mathbf{M}$ guarantees exponential stability of the curve of equilibria as a whole, hence the statement:

The curve of equilibria attracts the trajectories of almost all the PDS_t, and it is therefore possible for the curve to be reached for some instant $t \in [0, T]$ [223].

In practice, the Hurwitz condition of matrix $-\mathbf{M}$ is reached by establishing a low-interference regime through dynamic spectrum management combined with *ad hoc* routing [66] and [211].

4.2.4.1 Computer Experiment: Double-Layer Dynamics A computer experiment is presented to support theoretical underpinnings of this section. According to the IEEE 802.11a standard for wireless local area networks, 48 out of 64 subcarriers are dedicated to data transmission. A network of 120 users is considered in what follows and it is assumed that 48 subcarriers can be potentially available for data transmission. Initially, the network faces spectrum scarcity and users are not able to transmit with their maximum powers. The following sequence of events happens in the network:

- New users join the network.
- Some of the subcarriers are not available anymore for secondary usage.
- The network is perturbed close to its equilibrium state by randomly changing the interference gains, which occurs due to the mobility of users.
- More subcarriers are available for secondary usage.
- Then, some of the subcarriers are not available anymore for secondary usage.

Note that the interference gains were changed randomly due to user mobility as well as appearance and disappearance of users.

The average transmit power and the average data rate achieved by users after occurrence of each event are depicted in Figure 4.16. Power and data rate are plotted versus the number of iterations. Occurrences of events are shown by dashed vertical lines. As shown in the figure, the network deviates from the equilibrium point, when an event occurs. Starting from an initial state dictated by the event, the network moves toward a new equilibrium. In the diagram, 10 iterations were shown between 2 consequent events but the convergence is fast and in practice less iterations are required to reach a new point on the curve of equilibria from an arbitrary initial state, provided that the conditions for existence of a stable curve of equilibria are satisfied. Also, when the initial state dictated by a discrete event is not far from the achieved equilibrium (i.e., the network is perturbed around its equilibrium state), the state trajectory remains close to the equilibrium, which is the case for event 3 [211].

4.3 MARKET-DRIVEN COGNITIVE RADIO NETWORKS

Next, we discuss the second regime of networks, namely market-driven cognitive radio networks, which builds on the theory of supply chain networks [18] to present a model for the spectrum market. This regime includes three different tiers of decision

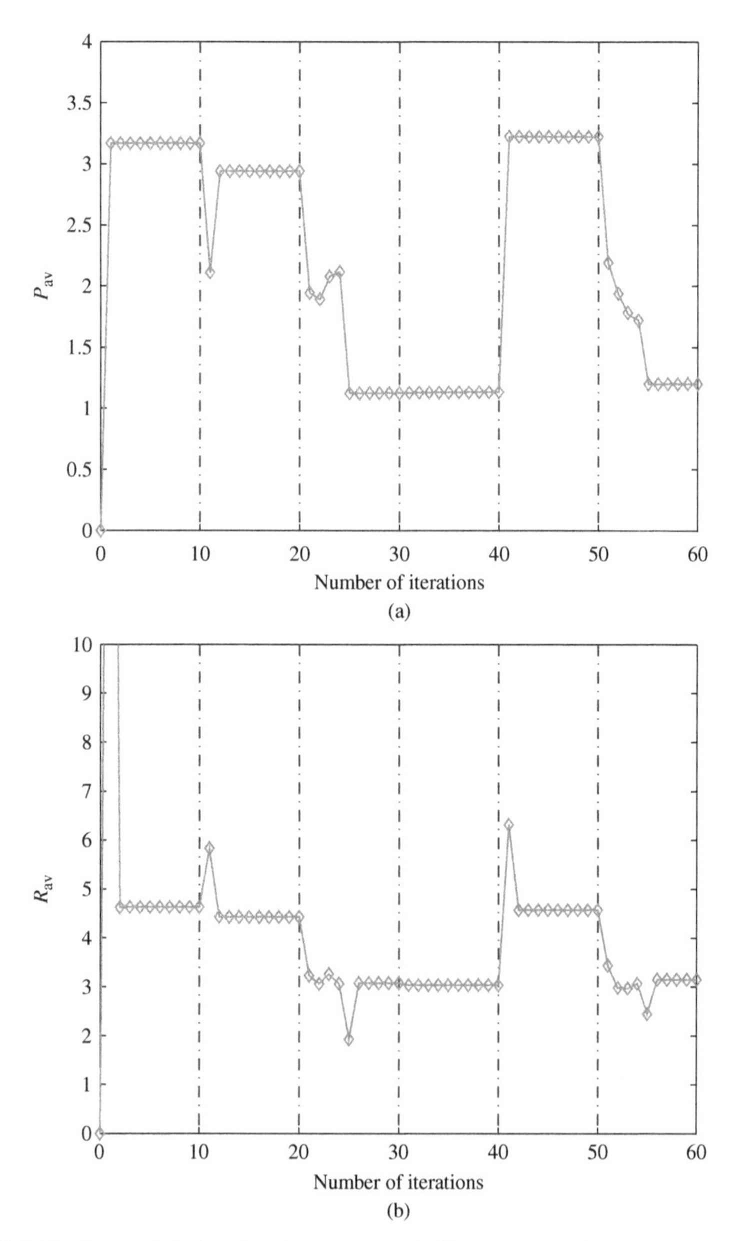

FIGURE 4.16 Dynamic behavior of a network of 120 users and 48 potentially available subcarriers. Dashed vertical lines show the occurrence of events. When an event occurs, network deviates from the established equilibrium. Starting from the initial state dictated by the event, network moves toward a new equilibrium: (a) average power and (b) average data rate are plotted versus the number of iterations. Source: Haykin and Setoodeh (2015) [37]. Reproduced with the permission of IEEE.

makers: spectrum legacy owners that provide service to primary users, secondary (cognitive radio) users, and so-called *spectrum brokers*, who mediate between primary and secondary users [37]. Figure 4.2 depicts the multilayered spectrum-supply chain network for this second regime [59]. In this new framework, spectrum brokers are responsible for assigning channels to cognitive radios through advertising the idle subbands within the legacy network; hence, the need for spectrum sensing is no longer a requirement; equally so, the need for dynamic spectrum manager is also eliminated. However, availability of subbands depends on the communication patterns of primary users in the region of interest, as it is with the open-access regime.

Furthermore, the scenario where spectrum holes come and go in an open-access regime, also applies to the market-driven regime. To be more specific, changes in the time horizon that involve the availability of spectrum holes do also arise in the market-driven regime. This, in turn, means that the spectrum brokers will have to cope with these changes. In other words, spectrum brokers play a key role in establishing a seamless communication process among the secondary users.

Continuing with the introductory material on the market-driven regime, the equilibrium point of the spectrum-supply chain network, in which none of the decision-makers have the incentive to change its policy unilaterally, is the solution of a respective VI problem. Also, the corresponding PDS, whose stationary points coincide with the equilibrium solutions of the VI, describes the transient behavior of the network.

There are two main sources of uncertainty in the spectrum-supply chain, which may lead to disruptions and associated risks: *demand-side risk* and *supply-side risk*. Secondary users are the cause of demand-side risk; they may freely join or leave the network and their communication patterns heavily depend on social events. Primary users are the cause of supply-side risk because appearance and disappearance of spectrum holes (underutilized subbands) depend on their communication patterns, which are also related to social events. Therefore, in a region, where there is a high demand for spectrum, the amount of idle spectrum available for secondary usage may be very limited. Since the impacts of supply chain disruptions may not remain locally and they may propagate globally all over the network, modeling and analysis methods that provide system-level views are essential for understanding the complex interactions of decision-makers. Such a global viewpoint will be helpful for risk management. In order to reduce the impacts of disruptions and associated risks, it is crucial to employ design methods that provide robustness from both local and global perspectives [86].

A three-dimensional resource space is considered, with time, frequency, and power as the relevant different dimensions. *Virtual power cube* (VPC), which is the minimum transmit power required to transmit an information unit, is used as the resource unit [235]. To explain, consider L spectrum legacy owners, each of which owns spectrum subbands including M_ℓ subcarriers ($\ell = 1, \dots, L$) and provides service to a number of primary users. Assume that there are B spectrum brokers and S secondary users. Each legacy owner tries to maximize its profit by leasing its idle and partially used subcarriers to spectrum brokers. Spectrum brokers, in turn, purchase the right for using those subcarriers from legacy owners and sell them to secondary users [236]. Spectrum brokers compete with each other in a noncooperative manner for gaining and trading that right. In the following subsections, the behavior of subsystems in each tier of the network as well as their optimality conditions are studied, and a VI formulation for finding the network equilibrium is presented.

4.3.1 Legacy Owners

Assume that legacy owner ℓ charges the spectrum broker b for subcarrier m_ℓ with the unit price $\rho^*_{m_\ell b}$. The optimal values of prices $\rho^*_{m_\ell b}$ for $\ell = 1, \ldots, L$; $b = 1, \ldots, B$; $m_\ell = 1_\ell, \ldots, M_\ell$; $(\sum_{\ell=1_\ell}^{L} M_\ell = M)$ are determined by finding the equilibrium point of the spectrum-supply chain network. Legacy owners compete in a noncooperative manner and each legacy owner ℓ solves the following optimization problem in order to maximize its own profit:

$$\begin{aligned} \max \quad & \sum_{m_\ell=1_\ell}^{M_\ell} \sum_{b=1}^{B} [\rho^*_{m_\ell b} p_{m_\ell b} - c_{m_\ell b}(p_{m_\ell b})] \\ & - \sum_{m_\ell=1_\ell}^{M_\ell} f_{m_\ell}(p_{m_\ell}) \\ \text{subject to}: \quad & \sum_{b=1}^{B} p_{m_\ell b} \le C_{m_\ell}, \forall m_\ell = 1_\ell, \ldots, M_\ell \\ & p_{m_\ell b} \ge 0, \forall m_\ell = 1_\ell, \ldots, M_\ell, \\ & \qquad \forall b = 1, \ldots, B. \end{aligned} \tag{4.56}$$

The first term in the objective function of the optimization problem (4.56) is the difference between revenue and transaction costs. *Transaction costs* are generally composed of two items [237]:

- *Coordination costs*, which are related to gathering, exchanging, and processing of information as well as incorporating the information into the decision-making process. To be more precise, these costs are associated with gaining *actionable information* [10].
- *Risk costs*, which are related to asset-specific investments and the possibility that other players involved in the transaction neglect their agreed upon responsibilities.

It may be difficult to find accurate models for transaction costs due to the fact that they are affected by the behaviors of players in the supply chain network.

The second term in the objective function of (4.56) is a penalty function, which reflects the effect of cross-channel interference as well as the effect of co-channel interference if legacy owners let the secondary users employ nonidle subcarriers in their frequency bands as well. It will also help to encourage secondary users to opt for less-utilized subcarriers. The respective transaction- and interference-cost functions, $c_{m_\ell b}(p_{m_\ell b})$ and $f_{m_\ell}(p_{m_\ell})$, are assumed to be continuously differentiable and convex. For instance, a second-order polynomial function can be used for $f_{m_\ell}(p_{m_\ell})$ and a second-order polynomial function without the constant term can be used for $c_{m_\ell b}(p_{m_\ell b})$. The first set of constraints guarantees that the *permissible interference power level* limit, C_{m_ℓ}, will not be violated in any subcarrier.

The Nash equilibrium of the noncooperative game between legacy owners, $(\mathbf{p}^{L*}, \mathbf{p}^{B*}) \in \mathcal{K}^L$, coincides with the solution of the following VI problem:

$$\sum_{\ell=1}^{L} \sum_{m_\ell=1_\ell}^{M_\ell} \sum_{b=1}^{B} \left[\frac{\partial c_{m_\ell b}(p^*_{m_\ell b})}{\partial p_{m_\ell b}} - \rho^*_{m_\ell b} \right] \times [p_{m_\ell b} - p^*_{m_\ell b}]$$
$$+ \sum_{\ell=1}^{L} \sum_{m_\ell=1_\ell}^{M_\ell} \frac{\partial f_{m_\ell}(p^*_{m_\ell})}{\partial p_{m_\ell}} \times [p_{m_\ell} - p^*_{m_\ell}] \geq 0. \quad (4.57)$$

$\forall (\mathbf{p}^L, \mathbf{p}^B) \in \mathcal{K}^L$, where $\mathbf{p}^L$ and $\mathbf{p}^B$ are, respectively, the LM- and LMB-dimensional power vectors, whose elements are p_{m_ℓ} and $p_{m_\ell b}$, with

$$\mathcal{K}^L = \{(\mathbf{p}^L, \mathbf{p}^B) | (\mathbf{p}^L, \mathbf{p}^B) \in \mathbb{R}_+^{LM+LMB};$$
$$\sum_{b=1}^{B} p_{m_\ell b} \leq C_{m_\ell}, \forall \ell = 1, \ldots, L, \forall m_\ell = 1_\ell, \ldots, M_\ell \}. \quad (4.58)$$

4.3.2 Spectrum Brokers

Spectrum brokers are involved in transactions with both legacy owners and secondary users. Assume that spectrum broker b charges the secondary user s with the unit price ρ^*_{bs}. This price is determined by finding the equilibrium point of the spectrum-supply chain network. Each spectrum broker b maximizes its own profit by solving the following optimization problem:

$$\max \quad \sum_{s=1}^{S} [\rho^*_{bs} p_{bs} - c_{bs}(p_{bs})] \quad (4.59)$$
$$- \sum_{m_\ell=1_\ell}^{M_\ell} \sum_{b=1}^{B} [\rho^*_{m_\ell b} p_{m_\ell b} + \hat{c}_{m_\ell b}(p_{m_\ell b})]$$
$$\text{subject to}: \sum_{s=1}^{S} p_{bs} = \sum_{\ell=1}^{L} \sum_{m_\ell=1_\ell}^{M_\ell} p_{m_\ell b}$$
$$p_{m_\ell b} \geq 0, \ \forall \ell = 1_\ell, \ldots, L; \quad \forall m_\ell = 1, \ldots, M_\ell$$
$$p_{bs} \geq 0, \ \forall s = 1, \ldots, S.$$

The objective function includes the revenue, payment to legacy owners, and the respective transaction costs. The first constraint expresses the fact that the total amounts of purchased and sold spectra are equal for each spectrum broker. It is also assumed that transaction costs $c_{bs}(p_{bs})$ and $\hat{c}_{m_\ell b}(p_{m_\ell b})$ are continuously differentiable and convex. For example, affine functions can be used for $c_{bs}(p_{bs})$ and $\hat{c}_{m_\ell b}(p_{m_\ell b})$.

The Nash equilibrium of the noncooperative game between spectrum brokers, $(\mathbf{p}^{B*}, \mathbf{p}^{S*}) \in \mathcal{K}^B$, coincides with the solution of the following VI problem:

$$\sum_{\ell=1}^{L} \sum_{m_\ell=1_\ell}^{M_\ell} \sum_{b=1}^{B} \left[\frac{\partial \hat{c}_{m_\ell b}(p^*_{m_\ell b})}{\partial p_{m_\ell b}} + \rho^*_{m_\ell b} \right] \times [p_{m_\ell b} - p^*_{m_\ell b}]$$
$$+ \sum_{b=1}^{B} \sum_{s=1}^{S} \left[\frac{\partial c_{bs}(p^*_{bs})}{\partial p_{bs}} - \rho^*_{bs} \right] \times [p_{bs} - p^*_{bs}] \geq 0. \tag{4.60}$$

$\forall (\mathbf{p}^B, \mathbf{p}^S) \in \mathcal{K}^B$, where $\mathbf{p}^B$ and $\mathbf{p}^S$ are, respectively, the *LMB*- and *BS*-dimensional power vectors, whose elements are $p_{m_\ell b}$ and p_{bs}, for which we have

$$\mathcal{K}^B = \{(\mathbf{p}^B, \mathbf{p}^S) | (\mathbf{p}^B, \mathbf{p}^S) \in \mathbb{R}_+^{LMB+BS};$$
$$\sum_{s=1}^{S} p_{bs} = \sum_{\ell=1}^{L} \sum_{m_\ell=1_\ell}^{M_\ell} p_{m_\ell b}, \ \forall b = 1, \ldots, B\}. \tag{4.61}$$

4.3.3 Secondary Users

Secondary users naturally compete with each other for resources also in a noncooperative manner. The secondary users' priority is to use license-free subbands. In unlicensed subbands, they compete for limited resources till the network reaches an equilibrium [66]. Since those subbands are usually too crowded, in the equilibrium condition some of the secondary users may not gain enough resources and they will have to lease the licensed subbands from legacy users. Each secondary user s tries to maximize its share of resources, d_s, subject to its power and budget constraints by solving the following optimization problem:

$$\max \quad d_s - \sum_{b=1}^{B} [\rho^*_{bs} + \hat{c}_{bs}] p_{bs} \tag{4.62}$$

$$\text{subject to}: \ d_s = \sum_{b=1}^{B} p_{bs}$$
$$\rho^*_{bs} + \hat{c}_{bs} \leq \frac{M_s^{\max}}{P_s^{\max}}$$
$$0 \leq d_s \leq P_s^{\max}$$
$$p_{bs} \geq 0, \ \forall b = 1, \ldots, B,$$

where $\hat{c}_{bs}$ is the unit transaction cost, $M_s^{\max}$ is the secondary user's maximum budget that can be dedicated to leasing spectrum, and $P_s^{\max}$ is the difference between the secondary user's power budget and the amount of transmit power that it uses in license-free subbands. The first inequality constraint set puts a limit on the price that the secondary user will pay for a resource unit.

The Nash equilibrium of the noncooperative game among the secondary users, $(\mathbf{p}^{S*}, \mathbf{d}^*) \in \mathcal{K}^S$, coincides with the solution of the following VI problem:

$$\sum_{b=1}^{B} \sum_{s=1}^{S} [\rho_{bs}^* + \hat{c}_{bs}] \times [p_{bs} - p_{bs}^*] - \sum_{s=1}^{S} [d_s - d_s^*] \geq 0. \tag{4.63}$$

$\forall (\mathbf{p}^S, \mathbf{d}) \in \mathcal{K}^S$, where $\mathbf{p}^S$ and $\mathbf{d}$ are, respectively, the BS- and S-dimensional vectors, whose elements are p_{bs} and d_s, for which we have

$$\begin{aligned} \mathcal{K}^S = \{(\mathbf{p}^S, \mathbf{d}) | (\mathbf{p}^S, \mathbf{d}) \in \mathbb{R}_+^{BS+S}; \\ \rho_{bs}^* + \hat{c}_{bs} \leq \frac{M_s^{\max}}{P_s^{\max}}, \forall b = 1, \ldots, B, \forall s = 1, \ldots, S; \\ d_s = \sum_{b=1}^{B} p_{bs} \leq P_s^{\max}, \ \forall s = 1, \ldots, S\}. \end{aligned} \tag{4.64}$$

4.3.4 Equilibrium of the Spectrum-Supply Chain Network

The network reaches an equilibrium point if the optimality conditions of decision-makers in the network including the legacy owners, the spectrum brokers, and the secondary users are all satisfied in a way that none of them has the incentive to unilaterally change its policy in accordance with the Nash equilibrium.

The equilibrium of the spectrum-supply chain network, $(\mathbf{p}^{L*}, \mathbf{p}^{B*}, \mathbf{p}^{S*}, \mathbf{d}^*) \in \mathcal{K}$, coincides with the solution of the following VI problem:

$$\begin{aligned} \sum_{\ell=1}^{L} \sum_{m_\ell=1_\ell}^{M_\ell} \frac{\partial f_{m_\ell}(p_{m_\ell}^*)}{\partial p_{m_\ell}} \times [p_{m_\ell} - p_{m_\ell}^*] - \sum_{s=1}^{S} [d_s - d_s^*] \\ + \sum_{b=1}^{B} \sum_{s=1}^{S} \left[\frac{\partial c_{bs}(p_{bs}^*)}{\partial p_{bs}} + \hat{c}_{bs} \right] \times [p_{bs} - p_{bs}^*] \\ + \sum_{\ell=1}^{L} \sum_{m_\ell=1_\ell}^{M_\ell} \sum_{b=1}^{B} \left[\frac{\partial c_{m_\ell b}(p_{m_\ell b}^*)}{\partial p_{m_\ell b}} + \frac{\partial \hat{c}_{m_\ell b}(p_{m_\ell b}^*)}{\partial p_{m_\ell b}} \right] \\ \times [p_{m_\ell b} - p_{m_\ell b}^*] \geq 0, \end{aligned}$$

$$\forall (\mathbf{p}^L, \mathbf{p}^B, \mathbf{p}^S, \mathbf{d}) \in \mathcal{K}, \tag{4.65}$$

where

$$\mathcal{K} = \Bigg\{ (\mathbf{p}^L, \mathbf{p}^B, \mathbf{p}^S, \mathbf{d}) | (\mathbf{p}^L, \mathbf{p}^B, \mathbf{p}^S, \mathbf{d}) \in \mathbb{R}_+^{(LM+S)(B+1)};$$

$$\sum_{b=1}^{B} p_{m_\ell b} \le C_{m_\ell}, \forall \ell = 1, \dots, L;\ \forall m_\ell = 1_\ell, \dots, M_\ell,$$

$$\sum_{s=1}^{S} p_{bs} = \sum_{\ell=1}^{L} \sum_{m_\ell=1_\ell}^{M_\ell} p_{m_\ell b},\ \forall b = 1, \dots, B;$$

$$d_s = \sum_{b=1}^{B} p_{bs} \le P_s^{\max},\ \forall s = 1, \dots, S;$$

$$\rho^*_{bs} + \hat{c}_{bs} \le \frac{M_s^{\max}}{P_s^{\max}}, \forall b = 1, \dots, B, \forall s = 1, \dots, S \Bigg\}. \tag{4.66}$$

The prices $\rho^*_{m_\ell b}$ can be recovered from (4.57) and (4.58) as

$$\rho^*_{m_\ell b} = \frac{\partial f_{m_\ell}(p^*_{m_\ell})}{\partial p_{m_\ell}} + \frac{\partial c_{m_\ell b}(p^*_{m_\ell b})}{\partial p_{m_\ell b}} \tag{4.67}$$

for any ℓ, m_ℓ, b such that $p^*_{m_\ell b} > 0$ and the prices ρ^*_{bs} can be recovered from (4.63) and (4.64) as

$$\rho^*_{bs} = \frac{M_s^{\max}}{P_s^{\max}} - \hat{c}_{bs} \tag{4.68}$$

for any b, s such that $p^*_{bs} > 0$.

The state vector of the spectrum-supply chain network is defined as

$$\mathbf{x} = (\mathbf{p}^L, \mathbf{p}^B, \mathbf{p}^S, \mathbf{d})^T. \tag{4.69}$$

By concatenating the corresponding terms in (4.65), we will get

$$\mathbf{F}(\mathbf{x}) = \begin{bmatrix} \left[\left[\frac{\partial f_{m_\ell}(p^*_{m_\ell})}{\partial p_{m_\ell}} \right]_{m_\ell=1_\ell}^{M_\ell} \right]_{\ell=1}^{L} \\ \left[\left[\left[\frac{\partial c_{m_\ell b}(p^*_{m_\ell b})}{\partial p_{m_\ell b}} + \frac{\partial \hat{c}_{m_\ell b}(p^*_{m_\ell b})}{\partial p_{m_\ell b}} \right]_{b=1}^{B} \right]_{m_\ell=1_\ell}^{M_\ell} \right]_{\ell=1}^{L} \\ \left[\left[\frac{\partial c_{bs}(p^*_{bs})}{\partial p_{bs}} + \hat{c}_{bs} \right]_{s=1}^{S} \right]_{b}^{B} \\ -\mathbf{1}_S \end{bmatrix}, \tag{4.70}$$

where $\mathbf{1}_S$ is an S-dimensional vector, whose elements are all 1. The VI problem in (4.65) and (4.66) can be rewritten in the compact form VI($\mathscr{K}, \mathbf{F}$).

The vector $\mathbf{x}^*$ is a Nash equilibrium point of the VI($\mathscr{K}, \mathbf{F}$) if, and only if, $\mathbf{x}^* \in \mathscr{K}$ and $\forall \mathbf{x} \in \mathscr{K}$, the following condition

$$(\mathbf{x} - \mathbf{x}^*)^T \mathbf{F}(\mathbf{x}^*) \geq 0 \tag{4.71}$$

is satisfied.

4.3.5 Network Dynamics

As with the open-access regime for cognitive radio networks, the PDS theory [221] is used to associate an ordinary differential equation (ODE) to the obtained VI problem, which allows us to study the equilibrium problem in a dynamic framework. The stationary points of the following PDS($\mathscr{K}, \mathbf{F}$)

$$\dot{\mathbf{x}} = \Pi_{\mathscr{K}}(\mathbf{x}, -\mathbf{F}(\mathbf{x})) \tag{4.72}$$

$$\mathbf{x}(t_0) = \mathbf{x}_0 \in \mathscr{K} \tag{4.73}$$

coincide with solutions of the VI($\mathscr{K}, \mathbf{F}$) problem of (4.71), where $\mathbf{F}$ and $\mathscr{K}$ are, respectively, described by (4.70) and (4.66).

4.3.6 Network Stability

Stability of the spectrum-supply chain network is guaranteed if $\mathbf{F}$ in the initial value problem described by equations (4.72) and (4.73), embodying a state-space model, is monotone [221].

4.3.7 The Transportation Network Representation of the Spectrum-Supply Chain Network

Suppose that there is a one-to-one correspondence between the nodes of two networks in such a way that the number of links between each pair of nodes in one network is equal to the number of links between the corresponding nodes in the other network; then, the two networks are said to be *isomorphic*. For directed networks, the corresponding links must have the same direction [238].

In [18], it is shown that the electric power supply chain network equilibrium model is isomorphic to a properly configured transportation network equilibrium model. Thus, adopting the approach of [18], we find that equivalence between the spectrum-supply chain network equilibrium model for market-driven regime and the transportation network equilibrium model is established [59].

For an illustrative example, the corresponding transportation network is depicted in Figure 4.17. It has five tiers with a single origin node 0 at the top tier and S destination nodes at the bottom tier. There are $1 + L + LM + B + S$ nodes, $L + LM + LMB + BS$ links, S origin/destination (O/D) pairs, and $LMBS$ paths. Consider the following provisions:

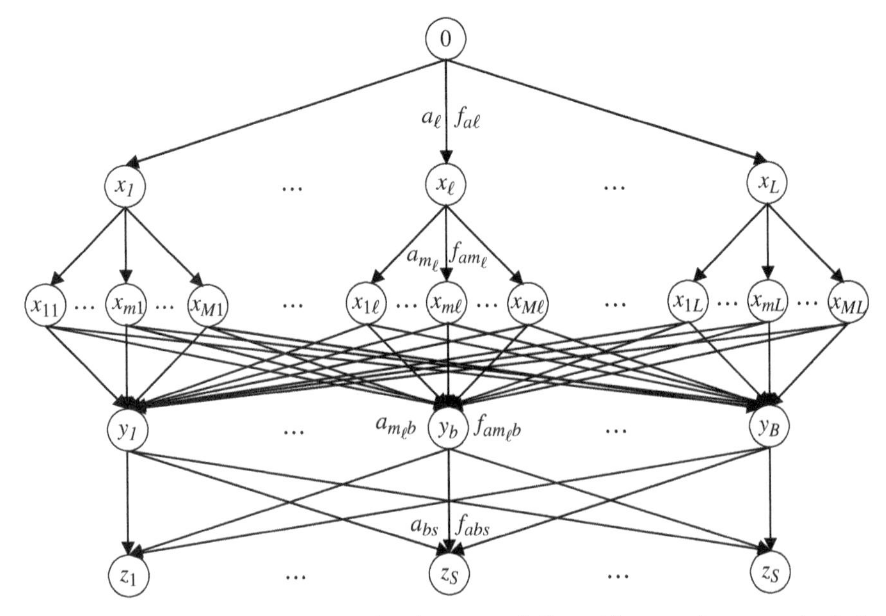

FIGURE 4.17 The transportation network representation of the spectrum-supply chain network for the market-driven regime. Source: Haykin and Setoodeh (2015) [37]. Reproduced with the permission of IEEE.

- a_ℓ denotes the link from node 0 to node x_ℓ with link flow f_{a_ℓ}, $\forall \ell = 1, \dots, L$.
- a_{m_ℓ} denotes the link from node x_ℓ to node x_{m_ℓ} with link flow $f_{a_{m_\ell}}$, $\forall \ell = 1, \dots, L$ and $\forall m_\ell = 1_\ell, \dots, M_\ell$.
- $a_{m_\ell b}$ denotes the link from node x_{m_ℓ} to node y_b with link flow $f_{a_{m_\ell b}}$, $\forall \ell = 1, \dots, L$, $\forall m_\ell = 1_\ell, \dots, M_\ell$, and $\forall b = 1, \dots, B$.
- a_{bs} denotes the link from node y_b to node z_s with link flow $f_{a_{bs}}$, $\forall b = 1, \dots, B$ and $\forall s = 1, \dots, S$.

Thus, a typical path, $q_{m_\ell bs}$, that connects the O/D pair $w_s = (0, z_s)$ consists of four links: $a_\ell, a_{m_\ell}, a_{m_\ell b}$, and a_{bs}. The associated flow on the path is $x_{q_{m_\ell bs}}$. Let d_{w_s} denote the demand associated with O/D pair w_s. Then, the following *conservation of flows* must be satisfied for the corresponding transportation network:

$$f_{a_\ell} = \sum_{m_\ell=1}^{M_\ell} \sum_{b=1}^{B} \sum_{s=1}^{S} x_{q_{m_\ell bs}}, \ \forall \ell = 1, \dots, L. \tag{4.74}$$

$$f_{a_{m_\ell}} = \sum_{b=1}^{B} \sum_{s=1}^{S} x_{q_{m_\ell bs}}, \ \forall \ell = 1, \dots, L, \quad \forall m_\ell = 1, \dots, M_\ell. \tag{4.75}$$

$$f_{a_{m_\ell b}} = \sum_{s=1}^{S} x_{q_{m_\ell bs}}, \ \forall \ell = 1, \ldots, L, \tag{4.76}$$

$$\forall m_\ell = 1, \ldots, M_\ell, \ \forall b = 1, \ldots, B.$$

$$f_{a_{bs}} = x_{q_{m_\ell bs}}, \ \forall b = 1, \ldots, B, \tag{4.77}$$

$$\forall s = 1, \ldots, S.$$

Also, we have

$$d_{w_s} = \sum_{\ell=1}^{L} \sum_{m_\ell=1}^{M_\ell} \sum_{b=1}^{B} x_{q_{m_\ell bs}}, \ \forall s = 1, \ldots, S. \tag{4.78}$$

We may thus identify the following three entities:

(i) A feasible path-flow pattern will be achieved if there are nonnegative path flows that satisfy equations (4.74)–(4.78).

(ii) A feasible path-flow pattern induces a feasible link-flow pattern, as described in [18].

(iii) A feasible link-flow pattern may be constructed, based on the corresponding feasible transmit power vectors in the spectrum-supply chain network model as shown by the following set of four equations:

$$\sum_{m_\ell=1}^{M_\ell} p_{m_\ell} \equiv f_{a_\ell} \tag{4.79}$$

$$\forall \ell = 1, \ldots, L.$$

$$p_{m_\ell} \equiv f_{a_{m_\ell}} \tag{4.80}$$

$$\forall \ell = 1, \ldots, L \quad \forall m_\ell = 1, \ldots, M_\ell.$$

$$p_{m_\ell b} \equiv f_{a_{m_\ell b}} \tag{4.81}$$

$$\forall \ell = 1, \ldots, L \quad \forall m_\ell = 1, \ldots, M_\ell \quad \forall b = 1, \ldots, B.$$

$$p_{bs} \equiv f_{a_{bs}} \tag{4.82}$$

$$\forall b = 1, \ldots, B \quad \forall s = 1, \ldots, S.$$

Building on the above-mentioned three entities, we may benefit from the literature on the transportation networks, and thereby deepen our knowledge of the underlying dynamics of market-driven cognitive radio networks [37].

4.4 SUPPLY CHAIN EFFICIENCY

As mentioned previously, the spectrum-supply chain network consists of multiple interacting and competitive decision-making units. These units are goal-seeking subsystems that have conflicting goals with their peers. Each unit may have minimum information about its competitors' budgets and policies as well as their capabilities for information gathering and processing. The decision-making units may choose a *pessimistic* attitude for dealing with this lack of knowledge about other players [66]. In the previous two sections, the multiple games that are played in different tiers of the spectrum-supply chain network for the two regimes were combined into a single game, which was formulated as a single variational inequality problem. A dynamic model for the network was derived that describes the disequilibrium behavior, and conditions for the existence of a unique equilibrium point were found.

An iterative algorithm based on time discretization of the PDS($\mathscr{K}$, $\mathbf{F}$) is proposed in [221] for computation of the network state trajectory. At each time-step t, the proposed algorithm solves the minimum-norm problem:

$$\min_{\mathbf{x}(t+1)\in\mathscr{K}} \quad \|\mathbf{x}(t+1) - [\mathbf{x}(t) - a(t)\mathbf{F}(\mathbf{x}(t))]\| \tag{4.83}$$

or equivalently, solves the following quadratic programming problem:

$$\begin{aligned}\min_{\mathbf{x}(t+1)\in\mathscr{K}} \quad & \frac{1}{2}\mathbf{x}^T(t+1)\mathbf{x}(t+1) \\ & -[\mathbf{x}(t) - a(t)\mathbf{F}(\mathbf{x}(t))] \cdot \mathbf{x}(t+1),\end{aligned} \tag{4.84}$$

where $\mathbf{x}$ denotes the state vector and "$\cdot$" signifies the dot product. A good approximation of the state trajectory may be achieved by choosing a small value for the step-size $a(t)$. Moreover, since the equilibrium is a point on the state trajectory, it can also be found by solving the above quadratic programming problem [66].

It is worth noting that even when a unique Nash equilibrium point exists, it may not be Pareto efficient. While anyone of the players does not have any incentive to unilaterally deviate from a Nash equilibrium solution, coalitions of groups of the players may find it appealing to do so. In other words, although a Nash equilibrium solution is immune to unilateral deviations by anyone of the players, it may be vulnerable to deviations by groups of the players that form some sort of coalitions [177].

Cooperative games in cognitive radio networks have been extensively studied for performance improvement. It has been suggested that groups of cognitive radios form coalitions for spectrum sensing [239], relaying [240], and collaboratively optimizing their transmit powers in a shared subband [63]. However, performance improvement in cooperative games comes at the cost of coordination through information exchange among players that form a coalition. Hence, generally speaking, sharing more information among players may lead to more efficient solutions. The trade-off between the efficiency and the amount of information exchange has attracted the attention of researchers in different disciplines. System designers, who deal with complex systems such as supply chain networks, have been looking for approaches that allow for converging to a Pareto-optimal set while maintaining the amount of information exchange somewhere between a noncooperative framework and a fully cooperative one [241].

In addition to the equilibrium solution efficiency, attention must be paid to the overall efficiency of the supply chain as well. In order to investigate the supply chain efficiency, two different levels of performance must be taken into account [242]:

- Performance of the supply chain components.
- Performance of the entire supply chain network.

The global performance of the supply chain depends on the local performances of the individual units as well as the trade-offs and interactions among them. It is known that selfish behavior of competing decision-making units may lead to inefficiency. Another cause for inefficiency would be if some of the legacy owners intentionally do not lease their idle subbands in order to increase the price and therefore their profits. This behavior is referred to as *rational inefficiency* and creates an *imperfectly competitive market* [243]. In general, different units may have different goals and it is unlikely that a single measure would be sufficient for evaluating the supply chain efficiency. The efficiency issue must be addressed by combining the payoff structure of multiple games associated with different tiers of the network into a single structure, hence the expression *game couplings* [244].

In summary, the spectrum-supply chain is a heterogeneous network in which decision-makers are different in their information-gathering and -processing capabilities, budgets, priorities, and policies. There are two ways to improve the efficiency of such a network [242]:

- Establishing coalition among groups of decision-makers via a proper information-sharing and message-passing mechanism among them [245].
- Emphasizing on a top-level management process.

The paradigm of *cognitive dynamic systems* [13] provides a good candidate for designing a supervisory system that plays the role of the top-level management [37].

4.5 CONCLUDING REMARKS

4.5.1 Two Regimes of Cognitive Radio Networks

In this chapter, we discussed the basics of two different regimes for cognitive radio networks: the open-access and the market-driven regimes. The first one embodies two kinds of users: primary (legacy) users and secondary (cognitive radio) users. On the other hand, the second one brings into play spectrum brokers for mediating between the primary and secondary users. In what follows, we summarize the functionalities of these two regimes [37]:

1. The *open-access regime*, the operation of which may be summarized as follows:
    - In so far as cognitive radio users are concerned, they rely on spectrum holes defined as unused subbands of the primary users (i.e., legacy owners) at a particular point in time and a particular location in space.

- Exploitation of the spectrum holes by the cognitive radio users has to be carried out in such a way that they must not affect the communication performance of the legacy users.
- For the discovery of spectrum holes, provisions must be made for the employment of spectrum sensing designed in such a way that secondary users are assured of a QoS comparable to that of the primary users, whenever and wherever open access to the secondary users is needed. Correspondingly, the spectrum sensors are required to be computationally efficient (i.e., fast) and robust in performance.
- The dynamic spectrum manager, responsible for the allocation of spectrum holes among competing cognitive radio users, has to operate in a *self-organized* manner; hence, the dynamic spectrum manager for secondary users may be viewed as the cognitive radio counterpart to the central base station in traditional wireless communications. The brain-inspired *self-organized dynamic spectrum management* (SO-DSM) fulfills the cognitive function just described [130].
- With QoS as a practical requirement, the transmit-power controller must be robust in having to face the nonstationary character of the wireless channel. The fact that spectrum holes come and go makes the channel even more nonstationary. The robust *iterative waterfilling algorithm* (*IWFA*) is well-suited for satisfying robustification of the transmit-power controller [66].
- Last, but by no means least, careful attention must be given to the underlying dynamics of the cognitive radio network so as to assure convergence of transient behavior to a satisfactory equilibrium. The generalized Nash equilibrium provides a suitable principle for the convergent behavior of cognitive radio network, although the Nash equilibrium may not be an optimal solution. There is a trade-off between optimality and ease of computation. In a cognitive radio network, which is a highly dynamic environment due to appearance and disappearance of both users and spectrum holes, finding a reasonably good solution (i.e., a suboptimal solution) that can be obtained fast enough is the only practical goal. Otherwise, spectrum holes may disappear before they can be utilized for communication [66].

2. The *market-driven regime* differs from the open-access regime by embodying *brokers* so as to mediate between the primary and secondary users, and assure similar QoS for both of them. Its operation may be summarized as follows:
    - A broker negotiates the price of an unused subband owned by a legacy owner; the agreed-on price must be acceptable to both the legacy owner and the secondary user. The net result of the negotiation is a "win–win" situation for both the legacy owner and the secondary user; the legacy owner wins by being paid for the unused subband, while the secondary user also wins by having access to the subband practically "on the fly."
    - In having the broker undertake this negotiation, functionality of cognitive radio is simplified by effectively dispensing with the following two cognitive functions: spectrum sensing and dynamic spectrum management. In other

words, spectrum sensing and dynamic spectrum management, essential to the open-access cognitive radio, are traded off for accommodating the employment of brokers in the market-driven cognitive radio. In so doing, a new functionality of brokers is brought into the structure of the market-driven regime.

- While the competition among open-access cognitive radio users is centered on unused spectrum holes (subbands), market-driven cognitive radio users compete for the "cheapest" prices through negotiation between legacy owners and secondary users via brokers embedded in the network.
- For a market-driven cognitive radio network to work successfully, the process of negotiations between primary and secondary users via brokers would have to be *efficient* (i.e., computationally fast) and *reliable* to satisfy a prescribed QoS.

4.5.2 Supply Chain Networks

As different as they are from each other, in actual fact open-access and market-driven cognitive radio networks are examples of supply chain networks, in which the spectrum plays the role of a commodity. Nevertheless, these two cognitive radio networks

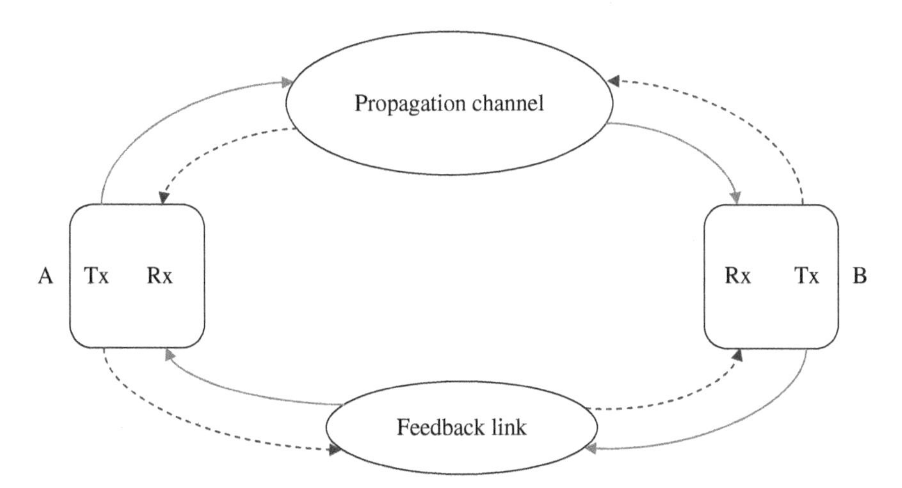

FIGURE 4.18 Directed information flow in the open-access regime, where two perception–action cycles operate in opposite directions, depending on which transceiver is listening and which one is speaking. With this scenario in mind, the Rx (i.e., the receiver) of transceiver B operates as the radio-scene analyzer, and the Tx (i.e., the transmitter) of transceiver A operates as the dynamic spectrum manager/transmit power controller, hence the perception–action cycle is in solid arrows. On the next perception–action cycle, the scenario is reversed for transceiver B, hence the cycle is in dashed arrows. And, so the cyclic directed information flow carries on. Source: Haykin and Setoodeh (2015) [37]. Reproduced with the permission of IEEE.

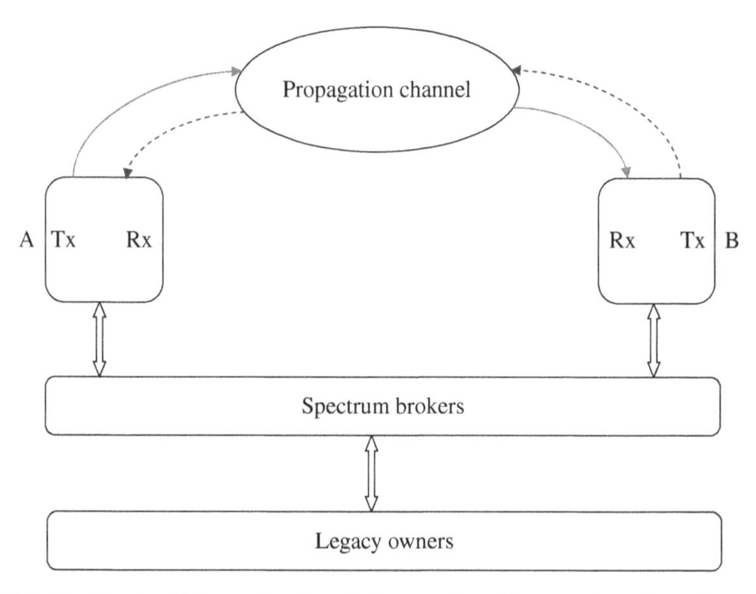

FIGURE 4.19 Directed information flow in the market-driven regime. Here, the propagation channel is available for communications after the negotiations involving the spectrum brokers are completed. The Rx (i.e., the receiver) of transceiver B links up with the Tx (i.e., the transmitter) of transceiver A via the combination of spectrum broker and legacy owner, thereby establishing a perception–action cycle as indicated by the solid arrows. For the next perception–action cycle, the directed information flow is reversed as shown by the dashed arrows. Source: Haykin and Setoodeh (2015) [37]. Reproduced with the permission of IEEE.

are radically different from each other, as illustrated in Figures 4.18 and 4.19, in the following two ways:

(i) The equilibrium and transient behavior of the open-access cognitive radio network are governed by the underlying dynamics of structure, where the networks of primary users and secondary users constitute a two-tier network.
(ii) On the other hand, the equilibrium and transient behavior of the market-driven cognitive radio network have an additional middle tier that accounts for the brokers, which makes the underlying mathematics much more complicated.

It is therefore natural for these two cognitive radio networks to behave entirely different from each other. Another way of differentiating between these two networks is to examine their *directed information flows*, as depicted in Figures 4.18 and 4.19.

4.5.3 Cognitive Radio Commercialization

The study of cognitive radio networks presented in this chapter would be incomplete without addressing where we stand on the commercialization of cognitive radio.

The emergence of cognitive radio started with the visionary paper by Mitola and Maguire [35]. In the course of past years since the publication of that paper, there is now an extensive literature on many facets of cognitive radio, albeit in a rather uneven manner. To be specific, a vast portion of that literature has focused on the cognitive function of spectrum sensing, yet the literature on the cognitive function of self-organized learning for spectrum management is meager in comparison. Needless to say, for a cognitive radio of the open-access regime to be commercializable, we would think that by now the literature on self-organized learning for dynamic spectrum management would be as rich as that of spectrum sensing. Alas, it is not!

Notwithstanding this imbalance of the literature on spectrum sensing and that on dynamic spectrum management, the logical commercializable space for cognitive radio networks of the open-access regime is *ad hoc* network applications as in the military, for example.

As for cognitive radio networks of the market-driven regime, they are well-suited for wireless communication network providers. In a way, this proposition would make commercial sense for the following reasons:

- Much, if not all, of the hardware and software of the network provider would remain essentially intact.
- The spectrum brokers and their software needs would be embedded into the network provider's structure, dispensing with spectrum sensing and dynamic spectrum management.
- With cognitive radio provisions built into a network provider, the QoS for cognitive radio users would be assured.

4.5.4 The Role of Cognition in Cognitive Radio Networks

The basic notion that distinguishes cognitive radio networks from ordinary wireless communications is *cognition*. As mentioned in the first chapter, there are five principles of cognition [13, 36]:

(i) *Perception–action cycle*, the function of which is to provide information gain about the radio environment, proceeding from one cycle to the next.
(ii) *Dynamic memory*, which is *predictive* in the sense that it predicts the consequences of actions taken by cognitive radio users.
(iii) *Attention*, the function of which is to improve the utilization of computational resources available to cognitive radio users.
(iv) *Intelligence*, which builds on the preceding three principles for optimal decision-making and control of data transmission by cognitive radio users.
(v) *Language* is more relevant in the context of a network of cognitive entities. For instance, a group of cognitive entities in the network may need a common language to dialogue among themselves for coordination. To this end, Mitola suggested the idea of a radio knowledge representation language (RKRL) for cognitive radio networks [29].

Principles (i), (iv), and (v) are common to both the open-access and market-driven regimes in their respective ways. However, it is in principles (ii) and (iii), where the two regimes of cognitive radio networks are different from each other [37]:

1. In the open-access regime, memory may be built into the radio-scene analyzer; also, memory does manifest itself in dynamic spectrum management [13].
2. In the market-driven regime, on the other hand, both memory and attention are accounted for by the spectrum broker, which is intuitively satisfying. If a human operator functions as the broker, then memory and attention are taken care of naturally. Equally so, the broker can be cognitized using neural computation [48], in which case, we have a *cognitive spectrum broker*, whose function is to mediate on behalf of legacy owners with secondary users, based on "machine-to-machine" communications.

In the open-access regime, spectrum sensing and dynamic spectrum management are at the heart of cognitive radio networks. On the other hand, in the market-driven regime, cognitive spectrum brokers are at the heart of cognitive radio networks.

As an alternative, in a database-oriented spectrum access scheme, a *radio environment map* (REM) plays the role of spectrum brokers as well as spectrum sensor and dynamic spectrum manager. Any kind of information about the network that can facilitate the coexistence of both primary and secondary users is stored in the REM. This includes but is not limited to physical environment, regulation, and policies as well as primary users' activity patterns and profiles. Secondary users consult this database to receive information about available subbands based on their geolocation (i.e., latitude and longitude). This way, a cognitive radio network can be formed from noncognitive wireless devices [246]. For instance, in TV white spaces, database-oriented spectrum access scheme would be very similar to the open-access regime except for the fact that radio transceivers do not need to be that advanced regarding their capabilities of sensing and processing. However, such databases require rapid updates [89].

Cognitive radio can be viewed as a radio with an internal computer or as a computer that communicates. Hence, it is a computer-intensive system [30]. In this regard, for resource allocation, attention should be paid to both radio and computation resources. In order to have more energy-efficient radios, energy-consuming computational tasks can be offloaded to close servers. Then, cognitive radio can use computation resources on demand via cloud computing [247]. This is quite similar to the idea that cognitive radio uses for spectrum utilization. In [248], it was suggested to build a cloud-based trading engine, through which secondary users can access radio resources in an auction-based competition. Such a cloud-based system can play the role of the spectrum brokers in future spectrum-supply chain networks.

It is worth noting that the market-driven spectrum-supply chain network can be approached in two different ways:

- One approach considers a network that involves a single tier of spectrum brokers, which was covered in this chapter.

- Another approach considers a network that involves two tiers of spectrum brokers that negotiate with each other, one on behalf of the network providers (i.e., legacy owners) and the other on behalf of the secondary users.

Both approaches have merits of their own. However, in rural areas the second approach might be preferred [249].

5

SUSTAINABILITY OF THE SPECTRUM-SUPPLY CHAIN NETWORK

One of the hallmarks of the fourth industrial revolution (Industry 4.0) is the ubiquitous connectivity of any device to any network, anytime at any place. However, bandwidth is the bottleneck engineering for all new technologies associated with Industry 4.0 such as cyber-physical systems, Internet of Things, Internet of Services, and cloud computing. Hence, the bandwidth issue must be taken into account in designing the 5G. Although all the proposed enabling technologies for 5G such as HetNets, millimeter waves, and massive MIMO have their own merits, when it comes to coping with the increasing demand for bandwidth in Industry 4.0, cognitive radio deserves special attention [250]. This chapter is focused on improving sustainability of the spectrum-supply chain network through building an artificial economy. It is known that Nash equilibrium is not immune to coalition formation. Hence, in order to guarantee the sustainability of communication network, a method for resource allocation is proposed to achieve Pareto optimality, which is based on the theory developed by the Swedish economist Eric Lindahl.

5.1 UNLICENSED BANDS AS PUBLIC GOODS

In economics, goods and services that are not accessible by consumers without paying a price are called *excludable*; otherwise, they are called *nonexcludable*. Another characteristic of interest for goods and services is their *rivalry* (subtractability), which

Fundamentals of Cognitive Radio, First Edition. Peyman Setoodeh and Simon Haykin.

deals with the question of whether access of a consumer will prevent or limit simultaneous access of another consumer. While excludability is a binary characteristic, rivalry is viewed as a continuum. To be more precise, goods or services may be regarded as nonrival only up to a certain capacity. For instance, coexistence of a number of radio transceivers in a certain band increases the level of interference that is experienced by each radio. In summary, goods and services are categorized into four groups in terms of excludability and rivalry [251]:

- Private goods are excludable and rivalrous.
- Common-pool resources are nonexcludable and rivalrous.
- Club goods are excludable and nonrivalrous.
- Public goods are nonexcludable and nonrivalrous.

In economic terms, licensed and unlicensed bands fall into the categories of private goods and common-pool resources, respectively. In this regard, spectrum sharing and allocation across licensed and unlicensed bands can be studied in a game-theoretic framework.

In a game-theoretic framework, usually type of players, their payoffs, and the way they interact with each other are formalized; then, the set of outcomes is searched for points with certain characteristics that match specific solution concepts. A popular solution concept is the celebrated *Nash equilibrium*, in which none of the players has the incentive to change its policy unilaterally. However, the Nash equilibrium achieved by a decentralized resource allocation approach may not be *Pareto optimal*. A feasible allocation is Pareto optimal if another feasible allocation does not exist, which is at least equally good for all users and better for some of them. In other words, in the state of Pareto optimality, a user's utility cannot be improved without making another user's utility worse. In noncooperative games, phenomena such as the tragedy of the commons may usually lead to inefficient equilibria [194].

As suggested in [37], radio transceivers, whether they belong to primary or secondary users, can be viewed as consumers in a spectrum-supply chain network. Designing a supervisory system that plays the role of the top-level management in cognitive radio networks will improve the performance. Following this way of thinking, the supervisory system can be used for *mechanism design* [68] in a game-theoretic framework. This way, the supervisory system allows for fixing the outcomes with desirable characteristics and then designing the rules of interaction among players in a way that desired outcomes are achieved as equilibria. Such outcomes must be well defined, feasible, and nonwasteful. Moreover, the mechanism should provide outcomes that are acceptable for players both in equilibrium and disequilibrium situations [252]. Building on the developed framework in [37], which views the wireless communication network as a spectrum-supply chain super-network, the supervisory system is used to design a market mechanism that yields *Lindahl* equilibria [253]. Lindahl allocations and prices are both individually rational and Pareto efficient in the sense of satisfying the *Samuelson* condition [254]. This condition states that the necessary condition for Pareto optimality is that the sum of every user's marginal rate of substitution of common-pool resources for private resources must equal the marginal cost of common-pool resources in terms of private resources [255].

5.2 THE SPECTRUM-SUPPLY CHAIN NETWORK AS AN ARTIFICIAL ECONOMY

In [37], two regimes were considered for spectrum sharing: the open-access regime and the market-driven regime. While the former is more appropriate for unlicensed bands, the latter is suitable for licensed bands and offers the incentive for network providers to support commercialization of cognitive radio. Regarding these two regimes, a theoretical framework was developed for harmonization between the two worlds of wireless communications (i.e., network providers and cognitive radio networks). In the proposed framework, the whole wireless world is viewed as a spectrum-supply chain network, in which spectrum has the role of commodity, network providers are suppliers of spectrum, and their primary users play the role of spectrum consumers. Another group of consumers, which are the secondary users, can join this supply chain network via spectrum brokers that play the role of mediators between network providers and secondary users. Spectrum brokers buy the right of using underutilized licensed subbands from network providers and sell it to secondary users. Such a supply chain is a three-tier network, where the decision-makers in each tier (i.e., network providers, spectrum brokers, and secondary users) compete against their peers to maximize their own profits. Secondary users have also the option of using unlicensed bands. They may try to gain as much spectrum as they can from unlicensed bands in a competitive manner. This competition occurs through indirect interaction among secondary users via the limited pool of resources (i.e., unlicensed bands) [66]. If the share they can get from unlicensed bands for free is not good enough for them, they can gain the rest of the required bandwidth by paying a price and leasing a portion of the licensed bands for a limited time.

From a game-theoretic perspective, secondary users are involved in two games: one game to get a share from unlicensed bands for free and the other game to get a share from licensed bands for the lowest price possible. Two other games are played in the spectrum-supply chain network: one among network providers and the other among spectrum brokers. In this framework, the prices that spectrum brokers pay to network providers and the prices that secondary users pay to spectrum brokers are considered as endogenous variables. A state-space model for the governing dynamics of the spectrum-supply chain network was derived in [37], whose stationary points are the Nash equilibria of the combined games in different tiers of the network. As explained in the previous chapter, this state-space model allows for analysis of both equilibrium and disequilibrium behaviors of the network. Regarding the fact that the network is not always in equilibrium, the state-space model is crucial for network analysis.

Network providers can temporarily share spectrum, infrastructure, or both for their mutual benefits. By the same token, cognitive radio networks can lease

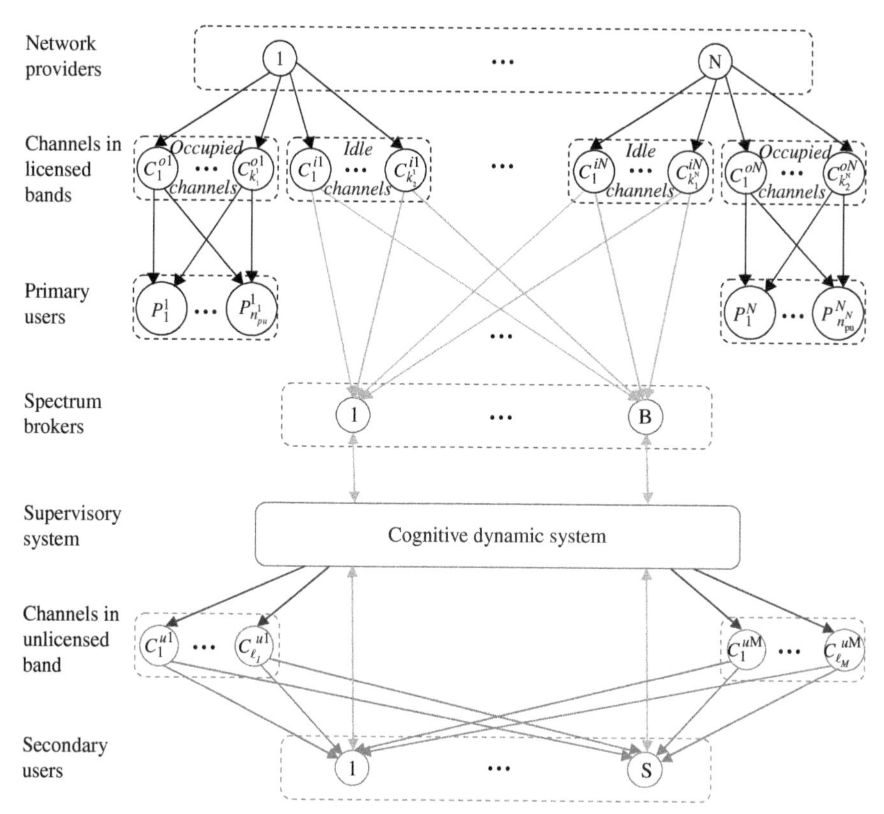

FIGURE 5.1 The spectrum-supply chain complex network facilitates spectrum sharing across licensed and unlicensed bands. The cognitive dynamic system in unlicensed bands is the counterpart of the network providers in licensed bands, and spectrum brokers harmonize the two wireless worlds. It is assumed that there are N network providers, where the jth network provider gives service to n_{pu}^{j} primary users and owns $k_1^j + k_2^j$ channels. Among the channels, k_1^j of them are occupied by the n_{pu}^{j} primary users that receive service from the jth network provider and the remaining k_2^j channels are assumed to be idle, which can be utilized for secondary usage. There are B spectrum brokers that buy the right of using the idle channels from network providers and sell it to the S active secondary users (i.e., cognitive radios). It is assumed that there are M unlicensed bands, where the kth band includes ℓ_k channels available to secondary users in an open-access regime.

spectrum, infrastructure, or both from network providers. A general economic model was presented in [256] that covers all of the mentioned cases of horizontal mergers of different networks. Merging and splitting of different networks within the spectrum-supply chain changes the topology of the supply chain and calls for redesigning the supply chain aimed at performance improvement. Hence, the structure of the spectrum-supply chain must be designed and redesigned in a dynamic manner [257]. With this framework in mind, the spectrum-supply chain network is indeed a complex network in its own right with dynamic topology. Moreover, the multitier structure of the network leads to complex multitimescale dynamics [59, 211].

It is worth noting that even when a unique Nash equilibrium point exists for spectrum sharing in the common pool of unlicensed bands, it may not be Pareto efficient. Moreover, information gathering (i.e., spectrum sensing) cost and constraints on computational time and power for finding suitable subbands that are less crowded must be taken into consideration. Regarding the limited sensing and computational ability of each user, some secondary users may have to form coalitions in order to be able to sweep a larger portion of the spectrum by dividing the associated costs between themselves. Through forming coalitions for spectrum sensing, secondary users decrease the chance of missing the hidden terminals as well [82]. On the other hand, spectrum brokers take care of these issues in licensed bands. Hence, although secondary users pay a price for using licensed bands, in those bands, the computational burden of spectrum sensing and dynamic spectrum management (i.e., channel allocation) will be lifted from their shoulders.

As an alternative to individually performing spectrum sensing and dynamic spectrum management by each cognitive radio, a database-oriented spectrum access scheme can be adopted and a *radio environment map* (REM) can be built [246]. In this framework, the supervisory system will be responsible for building the REM, keeping it updated, and act as a recommender system for resource allocation and distribution. This way, the supervisory system will be able to lead the network toward a solution, which is both individually and socially appealing. As mentioned before, such a solution concept is known as Lindahl equilibrium [253]. Figure 5.1 shows the proposed structure for the spectrum-supply chain network that facilitates spectrum sharing across licensed and unlicensed bands.

5.3 AIMING FOR LINDAHL EQUILIBRIA

Considering a three-dimensional resource space with time, frequency, and power as the relevant different dimensions, a resource unit is viewed as the minimum power required to transmit an information unit [37, 235]. Having a community of S active secondary users, it is assumed that there are m_ℓ licensed channels and m_u unlicensed channels that can potentially be used for communications. After paying the proper price, it is assumed that secondary user i gains access to a number of channels within the licensed bands. The licensed resources consumed by user i are represented by the vector $\boldsymbol{\phi}^i$. The unlicensed bands are accessible to every secondary user and the vector that represents the unlicensed resources consumed by the user i is denoted by the vector $\boldsymbol{\psi}^i$. User i's utility function is then denoted by $U^i(\boldsymbol{\phi}^i, \boldsymbol{\psi}^i)$. The m_ℓ-dimensional price vector for accessing licensed bands is denoted by $\mathbf{p}$ and the m_u-dimensional price vector that user i pays to access the REM is denoted by $\mathbf{q}^i$. Let us also define $\mathbf{q} = \sum_{i=1}^{S} \mathbf{q}^i$. This price includes both paying a fee for receiving information and/or contributing to forming and updating the map. Simply put, it is a combination of financial and informational (i.e., computational) costs. A characteristic of the Lindahl model, which may seem counterintuitive at the first sight, can be summarized as follows [258]:

- Different secondary users can use different numbers of licensed channels but at equilibrium, they pay the same price for a specific channel.
- All of the secondary users have access to the same unlicensed bands, but they may pay different prices, whether it be financially or computationally.

A few definitions are now recalled from [258]. The aggregate consumption of licensed and unlicensed resources is represented by the vectors $\boldsymbol{\phi} = \sum_{i=1}^{S} \boldsymbol{\phi}^i$ and $\boldsymbol{\psi} = \sum_{i=1}^{S} \boldsymbol{\psi}^i$, respectively. The vector $\boldsymbol{\phi}$ can also be viewed as the aggregate private resources that were supplied by network providers for secondary usage, which depends on the communication patterns of their primary customers. An aggregate output vector is defined by the concatenation of these two aggregate consumption vectors $[\boldsymbol{\phi}^T, \boldsymbol{\psi}^T]^T$. The set of feasible aggregate outputs $\mathscr{K} \subset \mathbb{R}^{m_\ell + m_u}$ is assumed to be closed and bounded. The set of feasible allocations includes all the allocations $\{\boldsymbol{\phi}^1, \ldots, \boldsymbol{\phi}^S, \boldsymbol{\psi}^1, \ldots, \boldsymbol{\psi}^S\}$ such that $[\sum_{i=1}^{S} (\boldsymbol{\phi}^i)^T, \sum_{i=1}^{S} (\boldsymbol{\psi}^i)^T]^T \in \mathscr{K}$. A wealth distribution function is considered for determining each user's wealth in terms of financial and computational budgets, $w^i(\mathbf{p}, \mathbf{q})$. It is assumed that the wealth function has the following property:

$$\sum_{i=1}^{S} w^i(\mathbf{p}, \mathbf{q}) = \max \ \mathbf{p}^T \boldsymbol{\phi} + \mathbf{q}^T \boldsymbol{\psi} \tag{5.1}$$

$$\text{subject to}: \quad \begin{bmatrix} \boldsymbol{\phi} \\ \boldsymbol{\psi} \end{bmatrix} \in \mathscr{K}.$$

Simply put, the cognitive radio network's wealth (i.e., the total wealth of all secondary users) will be equal to the most valuable feasible aggregate output for any licensed-channel prices and any unlicensed-channel Lindahl prices.

A Lindahl equilibrium consists of a vector of optimal prices for licensed channels, $\mathbf{p}^*$, each user's individualized optimal prices for unlicensed channels, known as Lindahl prices, $\{\mathbf{q}^{*1}, \ldots, \mathbf{q}^{*i}, \ldots, \mathbf{q}^{*S}\}$, and an optimal allocation of both licensed and unlicensed channels, $\{\boldsymbol{\phi}^{*1}, \ldots, \boldsymbol{\phi}^{*S}, \boldsymbol{\psi}^{*1}, \ldots, \boldsymbol{\psi}^{*S}\}$. User i's optimal resource consumption, $\{\boldsymbol{\phi}^{*i}, \boldsymbol{\psi}^{*i}\}$, is obtained by solving the following optimization problem:

$$\max_{\boldsymbol{\phi}^i, \boldsymbol{\psi}^i} \ U^i(\boldsymbol{\phi}^i, \boldsymbol{\psi}^i) \tag{5.2}$$

$$\text{subject to}: \quad (\mathbf{p}^*)^T \boldsymbol{\phi}^i + (\mathbf{q}^{*i})^T \boldsymbol{\psi}^i \leq w^i(\mathbf{p}^*, \mathbf{q}^*),$$

where $\mathbf{q}^* = \sum_{i=1}^{S} \mathbf{q}^{*i}$. The optimal price vector $\mathbf{p}^*$ is obtained as an endogenous variable, when the combined three-tier games between network providers, spectrum brokers, and secondary users in the licensed bands reach an equilibrium [37]. Lindahl equilibrium defined this way has the desired characteristic of Pareto optimality [255, 258]:

Theorem (Pareto Optimality of Lindahl Equilibrium): If all consumers have locally nonsatiated preferences, then a Lindahl equilibrium is Pareto optimal.

Proof: Assume that allocation $\{\boldsymbol{\phi}^{*1}, \ldots, \boldsymbol{\phi}^{*S}, \boldsymbol{\psi}^{*1}, \ldots, \boldsymbol{\psi}^{*S}\}$ is a Lindahl equilibrium with the corresponding private-good prices $\mathbf{p}^*$ and Lindahl prices $\{\mathbf{q}^{*1}, \ldots, \mathbf{q}^{*i}, \ldots, \mathbf{q}^{*S}\}$. If $\{\boldsymbol{\phi}^{1}, \ldots, \boldsymbol{\phi}^{S}, \boldsymbol{\psi}^{1}, \ldots, \boldsymbol{\psi}^{S}\}$ is an alternative allocation, which is Pareto superior w.r.t. $\{\boldsymbol{\phi}^{*1}, \ldots, \boldsymbol{\phi}^{*S}, \boldsymbol{\psi}^{*1}, \ldots, \boldsymbol{\psi}^{*S}\}$, then it must satisfy the following condition for all individuals:

$$U^i(\boldsymbol{\phi}^i, \boldsymbol{\psi}^i) \geq U^i(\boldsymbol{\phi}^{*i}, \boldsymbol{\psi}^{*i}) \tag{5.3}$$

with strict inequality for some user j. Regarding inequality (5.3) and the fact that $\{\boldsymbol{\phi}^i, \boldsymbol{\psi}^i\}$ is not the solution of the optimization problem (5.2) instead of $\{\boldsymbol{\phi}^{*i}, \boldsymbol{\psi}^{*i}\}$, we conclude that $\{\boldsymbol{\phi}^i, \boldsymbol{\psi}^i\}$ must violate the constraint in (5.2). In other words, we will have $(\mathbf{p}^*)^T\boldsymbol{\phi}^i + (\mathbf{q}^{*i})^T\boldsymbol{\psi}^i \geq w^i(\mathbf{p}^*, \mathbf{q}^*)$ with strict inequality for user j, who strictly prefers $\{\boldsymbol{\phi}^{1}, \ldots, \boldsymbol{\phi}^{S}, \boldsymbol{\psi}^{1}, \ldots, \boldsymbol{\psi}^{S}\}$ to $\{\boldsymbol{\phi}^{*1}, \ldots, \boldsymbol{\phi}^{*S}, \boldsymbol{\psi}^{*1}, \ldots, \boldsymbol{\psi}^{*S}\}$. Adding these inequalities for all users, we obtain

$$\sum_{i=1}^{S}((\mathbf{p}^*)^T\boldsymbol{\phi}^i + (\mathbf{q}^{*i})^T\boldsymbol{\psi}^i) > \sum_{i=1}^{S} w^i(\mathbf{p}^*, \mathbf{q}^*). \tag{5.4}$$

From the definition of the wealth function in (5.1), we have

$$\sum_{i=1}^{S} w^i(\mathbf{p}^*, \mathbf{q}^*) \geq (\mathbf{p}^*)^T\boldsymbol{\phi}' + (\mathbf{q}^*)^T\boldsymbol{\psi}' \tag{5.5}$$

for all $\begin{bmatrix}\boldsymbol{\phi}' \\ \boldsymbol{\psi}'\end{bmatrix} \in \mathscr{K}$. From (5.4) and (5.5), we will conclude that

$$\sum_{i=1}^{S}((\mathbf{p}^*)^T\boldsymbol{\phi}^i + (\mathbf{q}^{*i})^T\boldsymbol{\psi}^i) > (\mathbf{p}^*)^T\boldsymbol{\phi}' + (\mathbf{q}^*)^T\boldsymbol{\psi}' \tag{5.6}$$

for all $\begin{bmatrix}\boldsymbol{\phi}' \\ \boldsymbol{\psi}'\end{bmatrix} \in \mathscr{K}$. Rewriting the left-hand side of the above inequality in a compact form, we have

$$(\mathbf{p}^*)^T\boldsymbol{\phi} + (\mathbf{q}^*)^T\boldsymbol{\psi} > (\mathbf{p}^*)^T\boldsymbol{\phi}' + (\mathbf{q}^*)^T\boldsymbol{\psi}'. \tag{5.7}$$

Since $(\mathbf{p}^*)^T\boldsymbol{\phi} + (\mathbf{q}^*)^T\boldsymbol{\psi}$ is greater than $(\mathbf{p}^*)^T\boldsymbol{\phi}' + (\mathbf{q}^*)^T\boldsymbol{\psi}'$ for any feasible $\boldsymbol{\phi}'$ and $\boldsymbol{\psi}'$, $\boldsymbol{\phi}$ and $\boldsymbol{\psi}$ must represent an infeasible allocation: $\begin{bmatrix}\boldsymbol{\phi} \\ \boldsymbol{\psi}\end{bmatrix} \notin \mathscr{K}$. In other words, an allocation $\begin{bmatrix}\boldsymbol{\phi} \\ \boldsymbol{\psi}\end{bmatrix}$ that is Pareto-superior to $\begin{bmatrix}\boldsymbol{\phi}^* \\ \boldsymbol{\psi}^*\end{bmatrix}$ does not exist. This completes the proof by contradiction. ■

5.4 CONCLUDING REMARKS

The underlying objective in this chapter is to harmonize the ecosystem of licensed bands owned by network providers with unlicensed bands used by cognitive radio networks. The whole purpose of this harmonization is the improved utilization of the precious natural resource: the radio spectrum. This objective has been realized on three accounts:

1. To exploit the CDS to improve the efficient utilization of the spectrum in unlicensed bands and do so in a fair-minded manner.
2. To accommodate the needs of secondary users, they look to network providers for having access to licensed bands.
3. The essence of harmonization is twofold: first, for the secondary users, the issue is exchanging money for buying spectrum; second, for the case of network providers, the issue is to get money for allowing the use of spectrum. Finally, these two points in mind are indeed the role of spectrum brokers.

In light of the points just made, an artificial economy may be built with the goal of achieving a Nash equilibrium, which is also Pareto optimal. Aiming for such an equilibrium that is known as Lindahl equilibrium improves the sustainability of the spectrum-supply chain network.

6 COGNITIVE HETEROGENEOUS NETWORKS

Heterogeneous networks provide an innovative and intelligent, yet realistic and pragmatic approach for expanding mobile network capacity in order to cope with the ever-increasing mobile broadband traffic [259]. In this newly emerged communication network paradigm, low-power nodes such as femtocells and relay nodes are deployed within macrocells [260]. Moreover, a variety of radio access technologies, architectures, and transmission schemes may be employed in the network; hence the name heterogeneous [259]. The kind of hyper-densification of small cells proposed by the heterogeneous network (HetNet) paradigm can be considered as one of the key enabling technologies for *5G* [261, 262]. The other two important technologies are mmWave [263], which puts new spectrum into play, and massive MIMO [264], which enhances the spectral efficiency by improving the bits/s/Hz/node. At the network level, *network function virtualization* (NFV) and *software-defined networking* (SDN) are viewed as two technological trends toward *cloud-based networking*. 5G must support an extremely wide range of wireless devices and in effect therefore, by moving toward 5G, the heterogeneity of networks will significantly increase [261]. This chapter builds on the framework, which was proposed in [256].

6.1 HETEROGENEOUS NETWORKS

In HetNets, while macrocells are responsible for broad coverage, femtocells are placed at the coverage holes of the macrocells as well as close to the high-demand

Fundamentals of Cognitive Radio, First Edition. Peyman Setoodeh and Simon Haykin.

locations [265]. These femtocells will improve indoor coverage and reduce traffic congestion at macrocell base stations. However, femtocells may provide limited offloading capability due to the severe interference from macrocells but on the positive side, configuring femtocells with restricted access, allow them to provide a better quality of service (QoS) for a small number of users through allocating more resources to them. Also, since users can receive a relatively strong signal from a close femtocell base station, the total power consumption can be reduced by femtocell deployment. Basically, femtocells can be installed by a plug-and-play method and some of them do not even need a wired backhaul, which makes them really cost-effective. In addition to femtocells, relay nodes are deployed close to the cell-edge to improve coverage, network throughput, and capacity by enhancing channel conditions. They can also reduce power consumption and costs [260]. Interference management (both intracell and intercell) is a critical issue in HetNets and calls for careful resource partitioning [265]. Moreover, in hyperdense HetNets, challenges such as resource utilization, cell association, fairness, complexity, QoS, self-organization, and mobility management deserve special attention [260, 265–267]. *Green communication* is another issue of interest, which deals with environmental concerns and addresses energy efficiency in communication networks (i.e., energy consumed by each network component as well as the total energy consumed in the whole network) [268].

In addition to energy efficiency, enhancing *spectral efficiency* through spatiotemporal reuse of spectrum has been a major concern in recent years. It continues to be a key challenge for 5G as well. *Cognitive radio* is expected to play a key role in improving the efficiency of spectrum utilization [30]. Hence, it is not surprising that cognitive radio has attracted the attention of researchers in the context of HetNets [269–271]. However, cognitive radios must be adaptive and frequency agile radios that can operate in a wide range of frequencies and support multiple radio-access technologies. *Spectrum markets* can provide a driving force for more efficient spectrum allocations through secondary usage [272–274]. Such markets can be viewed as a *supply chain network*, in which the spectrum plays the role of a commodity or product [59]. This framework can be extended even further to view the whole communication network as a spectrum-supply chain [37]. Such a framework can capture the dynamics of spectrum sharing among different users in both unlicensed and licensed bands. In the most general scenario, a user's demand can be met by taking a share from unlicensed bands and leasing (i.e., buying the right for temporarily using of) some licensed subbands. Spectrum markets can be managed by *brokers*, who buy and sell the right of using idle subbands over different timescales. However, two different sets of brokers may be involved in the spectrum-supply chain, where one set of brokers negotiate on behalf of the legacy owners and the other set on behalf of the secondary users.

More efficient utilization of spectrum can be achieved through resource sharing among different networks. In this regard, resource allocation is viewed as collaborative problem solving by a set of networks. Collaborative problem solving calls for autonomous dynamic *reconfiguration* for distribution of information and skills as well as adjusting goals [275]. Resource allocation in communication networks is a distributed-constrained optimization problem. Taking a network-centric (i.e., centralized) approach may lead to an optimal solution that is computationally intractable. On the other hand, terminal-centric (i.e., decentralized) radio resource management schemes may lead to equilibria that are far from optimal solutions, which can be provided by a centralized approach. In other words, in decentralized approaches, phenomena such as *tragedy of the commons* usually lead to inefficient equilibria. In order to improve the quality of the reached equilibrium in heterogeneous wireless networks, the following two operations can be employed for finding a middle-way solution between the centralized and decentralized approaches [275]:

- *Agent melting* refers to forming a super agent from two or more agents by unifying the knowledge, goals, and skills.
- *Agent splitting* refers to the process of deconstruction of a super agent.

In the context of communication networks, agent melting can be interpreted as resource sharing among different networks (i.e., merging of networks). A *cognitive dynamic system* (CDS) [13] can play the role of a supervisor that decides when the resource sharing would be beneficial. Then, the CDS can improve the efficiency through controlled employment of the above two operations. It is worth noting that a cognitive dynamic system has five pillars: perception–action cycle, memory, attention, intelligence, and language.

Decoupling of infrastructure and spectrum will provide an additional degree of freedom that leads to a more fluid market and more efficient spectrum utilization [272]. In the economic model, which is built around such a decoupling, network owners, legacy owners, and operators can be separate entities. Infrastructure sharing may include both passive and active elements of the network. While active elements refer to antennas, base stations, radio-access networks, and core networks, passive elements refer to physical spaces, backhaul connections, power supplies, and so on. Mobile *virtual* network operators and *roaming* are also allowed in this framework. Resource (i.e., infrastructure and spectrum) sharing has significant advantages including coverage expansion, smooth handling of spatial and temporal demand fluctuations, improved spectral and energy efficiency as well as revenue benefits [261]. Regarding the fact that HetNets are multitier networks in nature, the resource sharing in such networks can be performed at different levels. Hence, integration of spectrum-supply chains associated with different communication networks is quite similar to the *horizontal merger* of corporations [276].

The novelty of the perspective presented here is the conceptualization of telecommunication firms as networks of their economic activities. Within this framework, strategic advantages of spectrum-supply chain network integration (i.e., *synergy*) can be evaluated quantitatively based on a system-optimization viewpoint. Strategic importance of controlling the supply chain as a whole calls for a holistic viewpoint. The holistic view that comes with formulating the optimization problem at the system level allows for capturing different criteria such as spectral efficiency, energy efficiency, environmental impacts, and risk management. In light of *network science*, mergers or even acquisitions can be formulated by adding a set of appropriate links with their associated costs to join the two networks that were originally separated [276]. Similarly, infrastructure sharing, spectrum sharing, or both can be performed through adding proper extra links that join different spectrum-supply chain networks.

6.2 HORIZONTAL MERGERS OF SPECTRUM-SUPPLY CHAIN NETWORKS

This section presents the supply chain network models before and after the horizontal mergers. For the simplicity of representation, two networks denoted by A and B are considered. These two networks are integrated into a single network after the merger. Four different cases are considered [276]:

- *Case 0*: This is the baseline case, where networks A and B are considered individually and are independent before the merger (Figure 6.1).
- *Case 1*: Networks A and B merge and only share the spectrum (Figure 6.2).
- *Case 2*: Networks A and B merge and only share the infrastructure (Figure 6.3).
- *Case 3*: Networks A and B merge and share both the spectrum and infrastructure (Figure 6.4).

The system-optimization problem associated with each case is derived in the following.

6.2.1 Premerger Status

As shown in Figure 6.1, before the merger, each network i (i = A,B) has n^i_N network nodes (i.e., infrastructure components), n^i_C channels, and provides service to n^i_U users. Symbols N, C, and U refer to nodes, channels, and users, respectively. The graph that represents network i is denoted by $\mathcal{G}_i = [\mathcal{N}_i, \mathcal{L}_i]$, where $\mathcal{N}_i$ refers to nodes and $\mathcal{L}_i$ refers to directed links associated with economic activities. In this case, the super network composed of the two networks A and B is represented by the graph $\mathcal{G}^0 = [\mathcal{N}^0, \mathcal{L}^0] = \bigcup_{i=A,B}[\mathcal{N}_i, \mathcal{L}_i]$, where the superscript 0 refers to the case 0. A link from a channel to a node shows that the channel can be accessed through that node by a set of users. Also, a link from a node to a user shows that the user can potentially access all the channels that are accessible via that node. Depending on the geographical location of a user, cell association and spectrum allocation determine the set of paths $P^0_{U^i_k}$ that connect the origin node i (i = A, B) to the destination node (i.e., user) U^i_k in the network. We define an indicator δ_{ap} such that $\delta_{ap} = 1$ if path p includes link a and $\delta_{ap} = 0$, otherwise. Let P^0 denote the set of all paths $P^0 = \bigcup_{i=A,B;k=1,\ldots,n^i_U} P^0_{U^i_k}$. The flow f_a on link a corresponds to the amount of accessed resource units (i.e., bits/s/Hz/node), which comes at the estimated cost of $\hat{c}(f_a)$. Similarly, path flows are denoted by x_p. Also, the capacity on link a is denoted by u_a. The objective of each network is to minimize the total cost while meeting all the users' demands denoted by $d_{U^i_k}$. Therefore, the following system-optimization problem must be solved in order

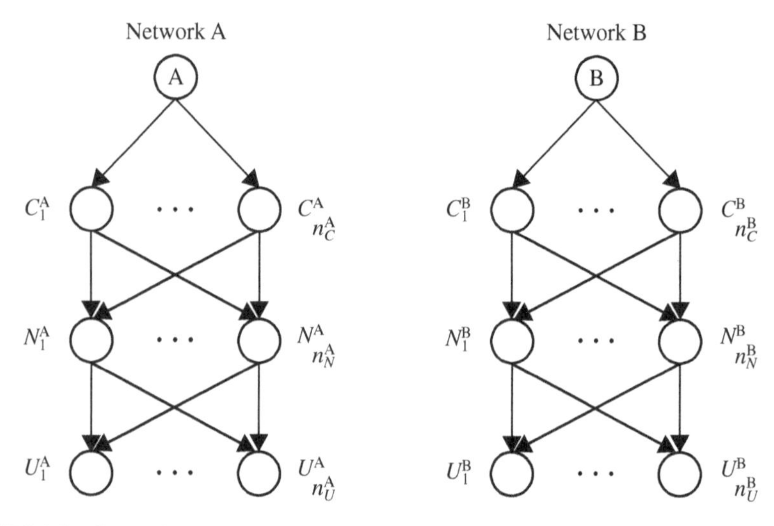

FIGURE 6.1 Case (0) Networks A and B prior to horizontal merger. Source: Haykin and Setoodeh (2015). Reproduced with the permission of IEEE.

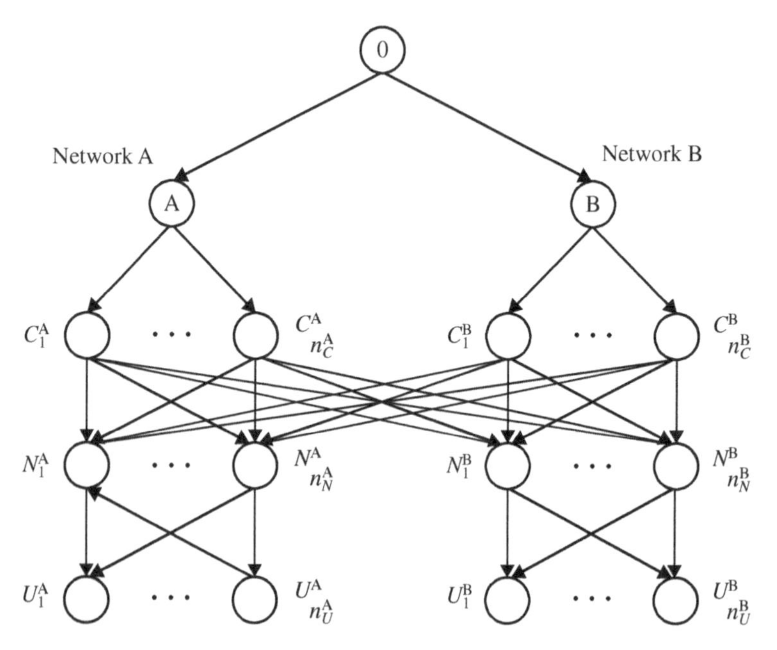

FIGURE 6.2 Case (1) Networks A and B merge: users associated with either network A or network B can now access each network's spectrum, but infrastructure of each original network deals with its original users. Source: Haykin and Setoodeh (2015). Reproduced with the permission of IEEE.

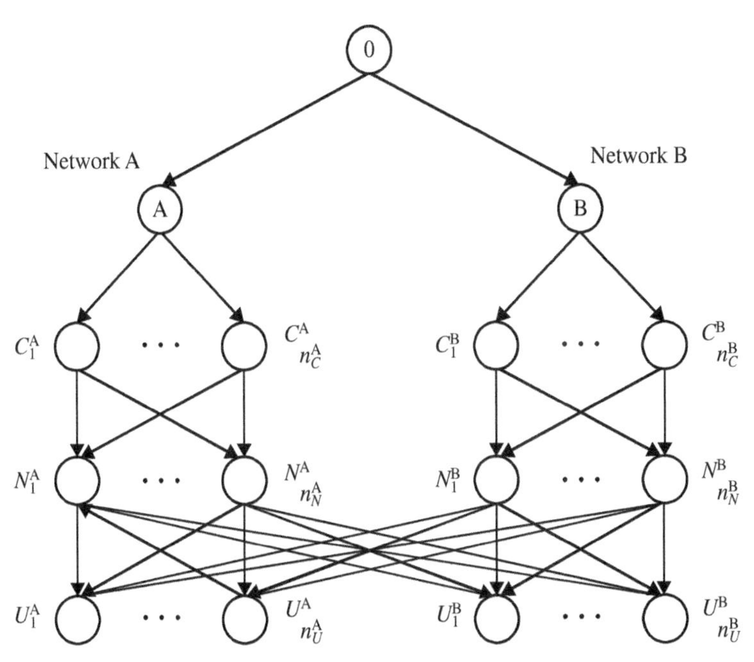

FIGURE 6.3 Case (2) users associated with either network A or network B can now access any network's infrastructure, but each user is supplied by each network's original spectrum. Source: Haykin and Setoodeh (2015). Reproduced with the permission of IEEE.

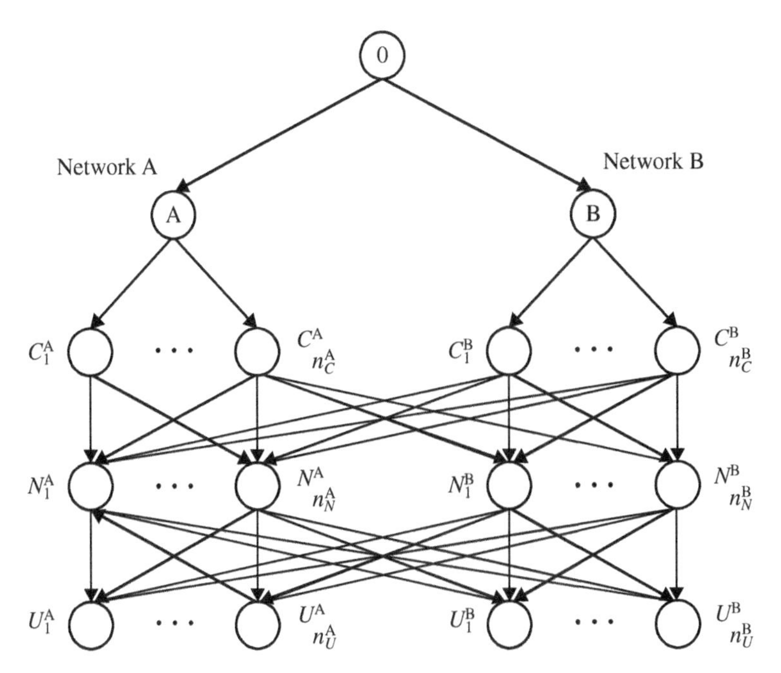

FIGURE 6.4 Case (3) networks A and B merge: users associated with either network A or network B can now access any network's infrastructure and spectrum. Source: Haykin and Setoodeh (2015). Reproduced with the permission of IEEE.

to minimize the total cost of the economic activities of both networks A and B, and yet meet the demands of all users that receive service from these two networks:

$$\begin{aligned}
&\underset{f_a}{\text{minimize}} \quad \sum_{a\in\mathscr{L}^0} \hat{c}(f_a) \qquad (6.1)\\
&\text{subject to:} \quad d_{U_k^i} = \sum_{p\in P^0_{U_k^i}} x_p, \; i = \text{A,B}; k = 1, \ldots, n_U^i\\
&f_a = \sum_{p\in P^0} \delta_{ap} x_p, \; \forall p \in P^0\\
&x_p \geq 0, \; \forall p \in P^0\\
&f_a \leq u_a, \; \forall a \in \mathscr{L}^0.
\end{aligned}$$

The first set of constraints guarantees that all users' demands will be met. The second set represents the conservation of flow, and states that the flow on a link must be equal to the sum of path flows for all paths, which include that link. The third set states that the path flows must be nonnegative and the fourth set guarantees some specified level of QoS by considering a capacity for each link flow.

In this case, two networks were considered before merger. Hence, there is no common link between the paths, which are associated with different networks. However, after merger, the number of paths, the sets of paths, the number of links, and the sets of links will all change. This point is further discussed in the following subsections.

6.2.2 Spectrum Sharing

Figure 6.2 depicts the resulting super network from merger of networks A and B, when they only share spectrum. In this case, the super node 0 and the corresponding links to nodes A and B symbolically represent the merger of the two networks. The graph that represents the super network is denoted by $\mathscr{G}^1 = [\mathscr{N}^1, \mathscr{L}^1]$, where $\mathscr{N}^1 = \mathscr{N}^0 \cup \text{node } 0$ and $\mathscr{L}^1 = \mathscr{L}^0 \cup \text{additional links due to spectrum sharing}$. The system-optimization problem that corresponds to this horizontal merger is formulated as follows:

$$\begin{aligned}
&\underset{f_a}{\text{minimize}} \quad \sum_{a\in\mathscr{L}^1} \hat{c}(f_a) \qquad (6.2)\\
&\text{subject to:} \quad d_{U_k^i} = \sum_{p\in P^1_{U_k^i}} x_p, \; i = \text{A,B}; k = 1, \ldots, n_U^i\\
&f_a = \sum_{p\in P^1} \delta_{ap} x_p, \; \forall p \in P^1\\
&x_p \geq 0, \; \forall p \in P^1\\
&f_a \leq u_a, \; \forall a \in \mathscr{L}^1.
\end{aligned}$$

6.2.3 Infrastructure Sharing

Figure 6.3 depicts the resulting super network from merger of networks A and B, when they only share infrastructure. In this case, the graph that represents the super network is denoted by $\mathcal{G}^2 = [\mathcal{N}^2, \mathcal{L}^2]$, where $\mathcal{N}^2 = \mathcal{N}^1$ and $\mathcal{L}^2 = \mathcal{L}^0 \cup$ additional links due to infrastructure sharing. The system-optimization problem that corresponds to this horizontal merger is formulated as follows:

$$
\begin{aligned}
\underset{f_a}{\text{minimize}} \quad & \sum_{a \in \mathcal{L}^2} \hat{c}(f_a) \qquad (6.3)\\
\text{subject to:} \quad & d_{U_k^i} = \sum_{p \in P^2_{U_k^i}} x_p, \; i = \text{A,B}; k = 1, \ldots, n_U^i \\
& f_a = \sum_{p \in P^2} \delta_{ap} x_p, \; \forall p \in P^2 \\
& x_p \geq 0, \; \forall p \in P^2 \\
& f_a \leq u_a, \; \forall a \in \mathcal{L}^2.
\end{aligned}
$$

6.2.4 Spectrum and Infrastructure Sharing

Figure 6.4 depicts the resulting super network from merger of networks A and B, when they share both the spectrum and infrastructure. In this case, which is the most general case of horizontal merger, the graph that represents the super network is denoted by $\mathcal{G}^3 = [\mathcal{N}^3, \mathcal{L}^3]$, where $\mathcal{N}^3 = \mathcal{N}^2 = \mathcal{N}^1$ and $\mathcal{L}^3 = \mathcal{L}^1 \cup \mathcal{L}^2$. The system-optimization problem that corresponds to this horizontal merger is formulated as follows:

$$
\begin{aligned}
\underset{f_a}{\text{minimize}} \quad & \sum_{a \in \mathcal{L}^3} \hat{c}(f_a) \qquad (6.4)\\
\text{subject to:} \quad & d_{U_k^i} = \sum_{p \in P^3_{U_k^i}} x_p, \; i = \text{A,B}; k = 1, \ldots, n_U^i \\
& f_a = \sum_{p \in P^3} \delta_{ap} x_p, \; \forall p \in P^3 \\
& x_p \geq 0, \; \forall p \in P^3 \\
& f_a \leq u_a, \; \forall a \in \mathcal{L}^3.
\end{aligned}
$$

6.3 SYNERGY MEASURE FOR HORIZONTAL MERGERS

The optimal set of flows, $f_a^*, \forall a \in \mathcal{L}^j$ $(j = 0, \ldots, 3)$ is obtained by solving the optimization problems (6.1) to (6.4). Having the optimal set of flows, the total minimal cost can be calculated as

$$\mathrm{TC}^j = \sum_{a \in \mathcal{L}^j} \hat{c}(f_a^*) \tag{6.5}$$

for $j = 0, \ldots, 3$ denoting different merger cases. Considering the premerger case as the baseline, the following *synergy measure* can be used to quantitatively evaluate the strategic advantage of different horizontal mergers [86]:

$$\mathcal{S}^j = [\frac{\mathrm{TC}^j - \mathrm{TC}^0}{\mathrm{TC}^0}] \times 100. \tag{6.6}$$

6.4 CONCLUDING REMARKS

Communication networks can be viewed as spectrum-supply chain networks. This framework provides a natural way to capture the hierarchical multitier structure of heterogeneous networks. Provisioning efficiency requirements of 5G, this chapter provided an economic model for resource sharing among different networks, where resources include both spectrum and infrastructure. The resource-sharing problem was considered as supply-chain integration or network horizontal merger, which can be conducted at different levels based on decoupling of spectrum and network infrastructure. The strategic advantage of mergers at different levels can be compared via a quantitative synergy measure, which considers the premerger status as the baseline.

The presented formulation is quite general and can be applied to any number of networks. It is built on a holistic viewpoint for optimally controlling the super network (i.e., network of networks), which is composed of different spectrum-supply chain networks. Also, the corresponding objective functions can capture different criteria to meet the expectations for 5G. These criteria include but not limited to green communication, spectral efficiency, and coverage expansion. The corresponding optimization problems can be reformulated as variational inequalities, which facilitate equilibrium computation and analysis.

Different networks can significantly benefit from supervisory entities that can be built via the cognitive dynamic systems paradigm. To be more precise, networks can be enabled with cognition at different levels (at the levels of components, subnetworks, and the whole network). The supervisory entities in a network will be responsible for making decisions whether to merge with another network and at what level. These merging and splitting decisions can be made locally (at the level of subnetworks) or globally (at the level of the whole network) according to the spatial and temporal demand fluctuations. This approach will pave the way for improving the quality of the equilibria of the resource-allocation problem by providing a middle-way solution between what can be found by centralized and decentralized techniques. This can be considered as an important advantage of the proposed framework. However, it requires further investigation.

Appendix A

MATHEMATICAL MODEL FOR OPEN-ACCESS COGNITIVE RADIO NETWORKS

This appendix derives the mathematical model presented for open-access cognitive radio networks. As mentioned before, the optimization problem in (3.82), which is repeated here for convenience,

$$\begin{aligned}
\max_{\mathbf{p}^i} \quad & f^i(\mathbf{p}^1, \dots, \mathbf{p}^n) = \sum_{k=1}^{m} \log_2\left(1 + \frac{p_k^i}{I_k^i}\right) \\
\text{subject to}: \quad & \sum_{k=1}^{m} p_k^i \le p_{\max}^i \\
& p_k^i + I_k^i \le \mathrm{CAP}_k, \ \forall k \notin PS \\
& p_k^i = 0, \ \forall k \in PS \\
& p_k^i \ge 0
\end{aligned} \tag{A.1}$$

can be rewritten as

$$\begin{aligned}
\min_{\mathbf{p}^i} \quad & -f^i(\mathbf{p}^1, \dots, \mathbf{p}^n) \\
\text{subject to}: \quad & \mathbf{p}^i \in \mathcal{K}^i,
\end{aligned} \tag{A.2}$$

where $\mathcal{K}^i$ is user i's feasible set. We recall the following theorem from [94, 221].

Fundamentals of Cognitive Radio, First Edition. Peyman Setoodeh and Simon Haykin.

Theorem A.1: Let $\mathscr{K}^i$ be a closed convex subset of $\mathbb{R}^n$ and $-f^i$ be a convex and continuously differentiable function in $\mathbf{p}^i$ for $i = 1, \ldots, n$. $\mathbf{p}^* = [\mathbf{p}^{*1^T}, \ldots, \mathbf{p}^{*n^T}]^T$ is an equilibrium of the game if it is a solution of the following VI problem VI($\mathscr{K}$, **F**):

$$(\mathbf{p} - \mathbf{p}^*)^T \mathbf{F}(\mathbf{p}^*) \geq 0, \tag{A.3}$$

where

$$\mathbf{F}(\mathbf{p}) = -[\nabla_{\mathbf{p}^i} f^i]_{i=1}^m \tag{A.4}$$

and

$$\mathscr{K} = \Bigg\{ \mathbf{p} \in \mathbb{R}^{n \times m} |\ p_k^i = 0,\ \forall k \in PS, \forall i = 1, \ldots, n;$$
$$0 \leq p_k^i + I_k^i \leq \text{CAP}_k,\ \forall k \notin PS, \forall i = 1, \ldots, n;$$
$$\sum_{k=1}^{n} p_k^i \leq p_{\max}^i,\ \forall i = 1, \ldots, n \Bigg\}. \tag{A.5}$$

Calculating the gradients in (A.4) leads to fractional terms with the sum of the power and interference plus noise in the denominators:

$$\nabla_{\mathbf{p}^i} f^i = \left[\frac{1}{p_1^i + I_1^i}, \ldots, \frac{1}{p_m^i + I_m^i} \right]^T \tag{A.6}$$
$$= \left[\frac{1}{\sigma_1^i + \sum_{j=1}^n \alpha_1^{ij} p_1^j}, \ldots, \frac{1}{\sigma_n^i + \sum_{j=1}^n \alpha_n^{ij} p_n^j} \right]^T$$

Building on [187], the proposed reformulation of IWFA was extended to the cognitive-radio problem in [66]. Here, the results reported in [66] will be further extended.

The *Lagrangian* of the optimization problem in (A.1) for user i is now written as

$$L^i(\mathbf{p}^1, \ldots, \mathbf{p}^n) = -f^i + u^i \left(\sum_{k=1}^{m} p_k^i - p_{\max}^i \right) \tag{A.7}$$
$$+ \sum_{k \notin PS} \gamma_k^i \left(\sigma_k^i + \sum_{j=1}^{n} \alpha_k^{ij} p_k^j - \text{CAP}_k \right) + \sum_{k \in PS} \lambda_k^i p_k^i.$$

Therefore, we have

$$\begin{cases} \gamma_k^i = 0, \lambda_k^i > 0 & k \in PS \\ \lambda_k^i = 0, \gamma_k^i > 0 & k \notin PS \end{cases} \tag{A.8}$$

The Karush–Kuhn–Tucker (KKT) conditions [277–279] for user i and $\forall k = 1, \ldots, m$ are as follows:

$$0 \le p_k^i \perp -\frac{1}{\sigma_k^i + \sum_{j=1}^n \alpha_k^{ij} p_k^j} + u^i + \gamma_k^i + \lambda_k^i \ge 0 \tag{A.9}$$

$$0 \le u^i \perp p_{\max}^i - \sum_{k=1}^m p_k^i \ge 0$$

$$0 \le \gamma_k^i \perp \text{CAP}_k - \sigma_k^i - \sum_{j=1}^n \alpha_k^{ij} p_k^j \ge 0, \ \forall k \notin PS$$

$$p_k^i = 0, \ \forall k \in PS,$$

where "$\perp$" signifies orthogonality of the corresponding variables.

Regarding the availability of spectrum for secondary usage, two cases may happen. If the network faces spectrum scarcity, some of the users may not be able to transmit with their maximum powers. Then, the first constraint in (A.1) will be redundant for those particular users. On the other hand, if the available spectrum is sufficient for all of the users to transmit with their maximum powers, the following inequality will be satisfied:

$$\sum_{j=1}^n p_{\max}^j < \sum_{k \notin PS} (\text{CAP}_k - \sigma_k^{\max}) \tag{A.10}$$

where $\sigma_k^{\max}$ is the maximum normalized background noise power on subcarrier k. In this case, similar to Proposition 1 of [187], which was proved for DSL, it can be shown that the system described in (A.9) is equivalent to a *mixed linear complementarity system* (mixture of a linear complementarity problem with a system of linear equations) [280].

In a *nonlinear complementarity problem* (NCP), the vector $\mathbf{x} \in \mathbb{R}^n$, should be found such that

$$\mathbf{x} \ge 0, \quad \mathbf{F}(\mathbf{x}) \ge 0, \quad \mathbf{x}^T \mathbf{F}(\mathbf{x}) \ge 0,$$

where $\mathbf{F}$ is a nonlinear mapping from $\mathbb{R}^n$ to $\mathbb{R}^n$. The problem will be a *linear complementarity problem* (LCP) if $\mathbf{F} = \mathbf{M}\mathbf{x} + \mathbf{q}$ for a matrix $\mathbf{M}$ and a vector $\mathbf{q}$ with appropriate dimensions [186].

Proposition A.1: Suppose that (A.10) holds, then the system (A.9) is equivalent to the following mixed linear complementarity system:

$$0 \le p_k^i \perp \sigma_k^i + \sum_{j=1}^n \alpha_k^{ij} p_k^j + v^i + \varphi_k^i + \varsigma_k^i \ge 0 \tag{A.11}$$

$$0 \leq \varphi_k^i \perp \text{CAP}_k - \sigma_k^i - \sum_{j=1}^{n} \alpha_k^{ij} p_k^j \geq 0, \ \forall k \notin PS$$

$$p_{\max}^i - \sum_{k=1}^{m} p_k^i = 0$$

$$p_k^i = 0, \ \forall k \in PS,$$

where

$$v^i = -\frac{1}{u^i} \tag{A.12}$$

$$\varphi_k^i = \frac{\gamma_k^i \left(\sigma_k^i + \sum_{j=1}^{n} \alpha_k^{ij} p_k^j\right)}{u^i}$$

$$\varsigma_k^i = \frac{\lambda_k^i \left(\sigma_k^i + \sum_{j=1}^{n} \alpha_k^{ij} p_k^j\right)}{u^i}$$

and

$$u^i = -\frac{1}{v^i} \tag{A.13}$$

$$\gamma_k^i = -\frac{\varphi_k^i}{v^i \left(\sigma_k^i + \sum_{j=1}^{n} \alpha_k^{ij} p_k^j\right)}$$

$$\lambda_k^i = -\frac{\varsigma_k^i}{v^i \left(\sigma_k^i + \sum_{j=1}^{n} \alpha_k^{ij} p_k^j\right)}.$$

Proof: Let $(p_k^i, u^i, \gamma_k^i, \lambda_k^i)$ satisfy (A.9) and assume that the complement set of PS is nonempty. Since power is nonnegative and $\sigma_k^i > 0$, $0 \leq \alpha_k^{ij} \leq 1$, we may write

$$\sigma_k^i + \sum_{j=1}^{n} \alpha_k^{ij} p_k^j > 0 \quad \forall k = 1, \ldots, n \tag{A.14}$$

It can be proved by contradiction that $u^i > 0$. To show this, we first note that if $u^i = 0$, hence

$$\gamma_k^i + \lambda_k^i \geq \frac{1}{\sigma_k^i + \sum_{j=1}^{n} \alpha_k^{ij} p_k^j} > 0 \quad \forall k = 1, \ldots, n \tag{A.15}$$

If $k \notin PS$, then $\lambda_k^i = 0$ and from (A.15) we must have $\gamma_k^i > 0$. Regarding the third complementarity condition in (A.9), $\gamma_k^i > 0$ leads to

$$\text{CAP}_k - \sigma_k^i - \sum_{j=1}^{n} \alpha_k^{ij} p_k^j = 0 \tag{A.16}$$

Therefore, we have

$$\mathrm{CAP}_k - \sigma_k^{\max} \leq \mathrm{CAP}_k - \sigma_k^i = \sum_{j=1}^{n} \alpha_k^{ij} p_k^j \tag{A.17}$$

Taking the summation over $k \notin PS$ from both sides of this equation leads to

$$\sum_{k \notin PS} (\mathrm{CAP}_k - \sigma_k^{\max}) \leq \sum_{k \notin PS} \sum_{j=1}^{n} \alpha_k^{ij} p_k^j \tag{A.18}$$

$p_k^i = 0, \forall k \in PS$ and $\forall i = 1, \dots, n$, so we have

$$\sum_{k \in PS} \sum_{j=1}^{n} \alpha_k^{ij} p_k^j = 0 \tag{A.19}$$

Therefore, we can rewrite (A.18) as

$$\begin{aligned} \sum_{k \notin PS} (\mathrm{CAP}_k - \sigma_k^{\max}) \\ &\leq \sum_{k \notin PS} \sum_{j=1}^{n} \alpha_k^{ij} p_k^j + \sum_{k \in PS} \sum_{j=1}^{n} \alpha_k^{ij} p_k^j \\ &= \sum_{k=1}^{m} \sum_{j=1}^{n} \alpha_k^{ij} p_k^j \end{aligned} \tag{A.20}$$

Since $0 \leq \alpha_k^{ij} \leq 1$, we have

$$\sum_{j=1}^{n} \alpha_k^{ij} p_k^j \leq \sum_{j=1}^{n} p_k^j \tag{A.21}$$

and therefore

$$\sum_{k \notin PS} (\mathrm{CAP}_k - \sigma_k^{\max}) \leq \sum_{k=1}^{m} \sum_{j=1}^{n} p_k^j \tag{A.22}$$

Changing the order of the two summations in the right-hand side of (A.22), we get

$$\sum_{k \notin PS} (\mathrm{CAP}_k - \sigma_k^{\max}) \leq \sum_{j=1}^{n} \sum_{k=1}^{m} p_k^j \tag{A.23}$$

From the first inequality constraint of (A.1), we know that

$$\sum_{k=1}^{m} p_k^j \leq p_{\max}^j \tag{A.24}$$

Thus,

$$\sum_{k \notin PS} (\text{CAP}_k - \sigma_k^{\max}) \leq \sum_{j=1}^{n} p_{\max}^j \tag{A.25}$$

which contradicts (A.10). Thus, $\forall k \notin PS$ and $\forall i = 1, \dots, n$, in addition to λ_k^i, γ_k^i must be zero too and we must therefore have $u^i > 0$ in order to satisfy the first complementary condition in (A.9). Defining the following variables:

$$v^i = -\frac{1}{u^i} \tag{A.26}$$

$$\varphi_k^i = \frac{\gamma_k^i \left(\sigma_k^i + \sum_{j=1}^n \alpha_k^{ij} p_k^j\right)}{u^i}$$

$$\varsigma_k^i = \frac{\lambda_k^i \left(\sigma_k^i + \sum_{j=1}^n \alpha_k^{ij} p_k^j\right)}{u^i}$$

we do get a solution to (A.11).

Conversely, assume that $(p_k^i, v^i, \varphi_k^i, \varsigma_k^i)$ satisfies (A.11); this time, we must have $v^i < 0$. Otherwise,

$$\sigma_k^i + \sum_{j=1}^{n} \alpha_k^{ij} p_k^j + v^i + \varphi_k^i + \varsigma_k^i > 0 \tag{A.27}$$

and then the first complementarity condition in (A.11) yields

$$p_k^i = 0, \ \forall k = 1, \dots, m \tag{A.28}$$

which contradicts the equality constraint in (A.11). Therefore, (A.9) holds by having

$$u^i = -\frac{1}{v^i} \tag{A.29}$$

$$\gamma_k^i = -\frac{\varphi_k^i}{v^i \left(\sigma_k^i + \sum_{j=1}^n \alpha_k^{ij} p_k^j\right)}$$

$$\lambda_k^i = -\frac{\varsigma_k^i}{v^i \left(\sigma_k^i + \sum_{j=1}^n \alpha_k^{ij} p_k^j\right)}$$

This completes the proof. ∎

While each user solves the above *mixed linear complementarity problem* (MLCP) with time-varying constraints, they should finally reach an equilibrium. The linear equation in (A.11) dictates that each user transmits with its maximum power, which leads to the worst-case interference condition. Intuitively it makes sense that each user transmits with its maximum power in order to achieve maximum data rate.

In the most general case, where (A.10) is not valid, some of the users in the network will be able to transmit with their maximum powers and the others will not. We define two sets, $\mathcal{N}_1$ and $\mathcal{N}_2$, which include these two groups of users, respectively. Intuitively speaking, when users adjust their power vectors based on rate-adaptive waterfilling (A.1) in which they try to maximize their data rates subject to power constraints, they either transmit with their maximum power or with the highest power permitted by the interference limits. In the case of spectrum scarcity, where (A.10) is not valid, for user $i \in \mathcal{N}_1$, which is able to transmit with its maximum power, $u^i > 0$ and we have the following:

Proposition A.2: Suppose that (A.10) is not valid and user i is able to transmit with its maximum power, then the system (A.9) is equivalent to the mixed linear complementarity system (A.11).

Proof: The proof is straightforward. The same steps in the proof of *Proposition A.1* after showing that $u^i > 0$ should be followed. The relations between the corresponding variables were defined in (A.26) and (A.29). ■

On the other hand, when user i cannot transmit with its maximum power, the first constraint in (A.1) will be redundant and $u^i = 0$. The KKT conditions in (A.9) are reduced to

$$
\begin{aligned}
&0 \le p_k^i \perp -\frac{1}{\sigma_k^i + \sum_{j=1}^{n} \alpha_k^{ij} p_k^j} + \gamma_k^i + \lambda_k^i \ge 0 \qquad \text{(A.30)}\\
&0 \le \gamma_k^i \perp \mathrm{CAP}_k - \sigma_k^i - \sum_{j=1}^{n} \alpha_k^{ij} p_k^j \ge 0, \ \forall k \notin PS\\
&p_k^i = 0, \ \forall k \in PS
\end{aligned}
$$

In this case, we have

Proposition A.3: Suppose that (A.10) is not valid and the first constraint in (A.1) can be relaxed for user i, then the system (A.30) is equivalent to the following mixed linear complementarity system:

$$
\begin{aligned}
&0 \le p_k^i \perp \sigma_k^i + \sum_{j=1}^{n} \alpha_k^{ij} p_k^j + \varphi_k^i + \varsigma_k^i \ge 0 \qquad \text{(A.31)}\\
&\sigma_k^i + \sum_{j=1}^{n} \alpha_k^{ij} p_k^j = \mathrm{CAP}_k, \ \forall k \notin PS\\
&p_k^i = 0, \ \forall k \in PS
\end{aligned}
$$

where

$$\varphi_k^i = -\frac{1}{\gamma_k^i} \tag{A.32}$$

$$\varsigma_k^i = \frac{\lambda_k^i \left(\sigma_k^i + \sum_{j=1}^n \alpha_k^{ij} p_k^j\right)}{\gamma_k^i}$$

and

$$\gamma_k^i = -\frac{1}{\varphi_k^i} \tag{A.33}$$

$$\lambda_k^i = -\frac{\varsigma_k^i}{\varphi_k^i \left(\sigma_k^i + \sum_{j=1}^n \alpha_k^{ij} p_k^j\right)}$$

Proof: Let $(p_k^i, \gamma_k^i, \lambda_k^i)$ satisfy (A.30) and assume that the complement set of PS is nonempty. Since power is nonnegative and $\sigma_k^i > 0$, $0 \leq \alpha_k^{ij} \leq 1$, we may write

$$\sigma_k^i + \sum_{j=1}^n \alpha_k^{ij} p_k^j > 0 \quad \forall k = 1, \ldots, n \tag{A.34}$$

If $k \notin PS$, then $\lambda_k^i = 0$ and from the first complementary condition in (A.30) we have

$$\gamma_k^i \geq \frac{1}{\sigma_k^i + \sum_{j=1}^n \alpha_k^{ij} p_k^j} > 0 \quad \forall k = 1, \ldots, m \tag{A.35}$$

Regarding the second complementarity condition in (A.30), $\gamma_k^i > 0$ leads to

$$\sigma_k^i + \sum_{j=1}^n \alpha_k^{ij} p_k^j = \text{CAP}_k \tag{A.36}$$

Defining the following variables:

$$\varphi_k^i = -\frac{1}{\gamma_k^i} \tag{A.37}$$

$$\varsigma_k^i = \frac{\lambda_k^i \left(\sigma_k^i + \sum_{j=1}^n \alpha_k^{ij} p_k^j\right)}{\gamma_k^i}$$

we do get a solution to (A.31).

Conversely, assume that $(p_k^i, \varphi_k^i, \varsigma_k^i)$ satisfies (A.31). This time, we must have $\varphi_k^i < 0$. Otherwise,

$$\sigma_k^i + \sum_{j=1}^{m} \alpha_k^{ij} p_k^j + \varphi_k^i + \varsigma_k^i > 0. \tag{A.38}$$

Hence, the first complementarity condition in (A.31) yields

$$p_k^i = 0, \ \forall k = 1, \ldots, m \tag{A.39}$$

which contradicts the equality constraint in (A.31). Therefore, (A.30) holds by having

$$\gamma_k^i = -\frac{1}{\varphi_k^i} \tag{A.40}$$

$$\lambda_k^i = -\frac{\varsigma_k^i}{\varphi_k^i \left(\sigma_k^i + \sum_{j=1}^{n} \alpha_k^{ij} p_k^j \right)}.$$

This completes the proof. ∎

The linear equation in (A.31) suggests that, when user i cannot transmit with its maximum power, it transmits with the highest permissible power, dictated by the interference temperature limit. Again, intuitively it makes sense.

Users that belong to $\mathcal{N}_1$ solve the MLCP (A.11) and users that belong to $\mathcal{N}_2$ solve the MLCP (A.31). After concatenating the corresponding variables according to (4.2)–(4.4), and

$$\mathbf{CAP} = [CAP_k]_{k=1}^{m} = \left[\mathrm{CAP}_1 \cdots \mathrm{CAP}_m\right]^T, \tag{A.41}$$

the joint feasible set (A.5) can be rewritten as

$$\mathcal{K} = \Bigg\{ \mathbf{p} \in \mathbb{R}^{n\times m} | \ p_k^i = 0, \ \forall k \in PS, \forall i = 1, \ldots, n; \\ \sigma + \mathbf{M}\mathbf{p} \leq \mathbf{CAP}; \ \sum_{k=1}^{m} p_k^i \leq p_{\max}^i, \ \forall i = 1, \ldots, n \Bigg\}. \tag{A.42}$$

The MLCPs (A.11) and (A.31) are the KKT conditions for an affine variational inequality problem [94, 187], defined by the affine mapping

$$\mathbf{F}(\mathbf{p}) = \sigma + \mathbf{M}\mathbf{p} \tag{A.43}$$

and the polyhedron

$$\begin{aligned}\mathscr{K} = \{\mathbf{p} \in \mathbb{R}^{n\times m} | \ p_k^i = 0, \ \forall k \in PS, \forall i = 1, \dots, n; \\ p_k^i + I_k^i \leq \mathrm{CAP}_k, \ \sum_{k=1}^{m} p_k^i = p_{\max}^i, \ \forall k \notin PS, \ \forall i \in \mathscr{N}_1; \\ p_k^i + I_k^i = \mathrm{CAP}_k, \ \forall k \notin PS, \ \forall i \in \mathscr{N}_2\}\end{aligned} \tag{A.44}$$

Hence, the IWFA can be formulated as an AVI problem $\mathrm{VI}(K, \boldsymbol{\sigma} + \mathbf{Mp})$ or $\mathrm{AVI}(K, \boldsymbol{\sigma}, \mathbf{M})$.

Appendix B

PROOF OF THEOREMS

This appendix provides sketch of proof for theorems that were used either implicitly or explicitly in this chapter.

Theorem (VI Reformulation of a Nash Equilibrium Game): Let $\mathscr{K}^i$ be a closed convex subset of $\mathbb{R}^n$ and $-f^i$ be a convex and continuously differentiable function in $\mathbf{p}^i$ for $i = 1, \ldots, n$. $\mathbf{p}^* = \left[\mathbf{p}^{*1^T}, \ldots, \mathbf{p}^{*n^T}\right]^T$ is an equilibrium of the game if it is a solution of the following VI problem VI($\mathscr{K}$, **F**):

$$(\mathbf{p} - \mathbf{p}^*)^T \mathbf{F}(\mathbf{p}^*) \geq 0, \tag{B.1}$$

where

$$\mathbf{F}(\mathbf{p}) = -[\nabla_{\mathbf{p}^i} f^i]_{i=1}^m \tag{B.2}$$

and

$$\begin{aligned}\mathscr{K} = \Bigg\{ &\mathbf{p} \in \mathbb{R}^{n\times m} | \; p_k^i = 0, \; \forall k \in PS, \forall i = 1, \ldots, n; \\ &0 \leq p_k^i + I_k^i \leq \text{CAP}_k, \; \forall k \notin PS, \forall i = 1, \ldots, n; \\ &\sum_{k=1}^n p_k^i \leq p_{\max}^i, \; \forall i = 1, \ldots, n \Bigg\}.\end{aligned} \tag{B.3}$$

Proof: Proof of this theorem can be found in [94], Chapter 1, Proposition 1.4.2, and the discussion that follows it. Essential outline of the proof is as follows.

Fundamentals of Cognitive Radio, First Edition. Peyman Setoodeh and Simon Haykin.

The VI formulation of the game is obtained by writing down the KKT conditions for each player's optimization problem and concatenating the KKT systems of all players in the form of a mixed complementarity problem.

Due to convexity and minimum principle, $\mathbf{p}^*$ is a Nash equilibrium if, and only if, for each $i = 1, \ldots, m$

$$-(\mathbf{p}^i - \mathbf{p}^{i*})^T \nabla_{\mathbf{p}^i} f^i(\mathbf{p}^*) \geq 0, \ \forall \mathbf{p}^i \in K^i \tag{B.4}$$

Thus, if $\mathbf{p}^*$ is a Nash equilibrium, then by concatenating these individual VIs, it follows easily that $\mathbf{p}^*$ must solve the prescribed VI.

Conversely, if $\mathbf{p}^*$ solves the VI problem, then

$$(\mathbf{p} - \mathbf{p}^*)^T \mathbf{F}(\mathbf{p}^*) \geq 0, \ \forall \mathbf{p} \in K \tag{B.5}$$

In particular, for each $i = 1, \ldots, m$, let $\mathbf{p}$ be the vector whose jth subvector is equal to $\mathbf{p}^{*j}$ for $j \neq i$ and ith subvector is equal to $\mathbf{p}^i$, where $\mathbf{p}^i$ is an arbitrary element of the set K^i. The above inequality then becomes (B.4). ■

Theorem (Existence and Uniqueness of VI Solution): Let $\mathcal{K} \subseteq \mathbb{R}^n$ be closed convex and $\mathbf{F} : \mathcal{K} \subseteq \mathbb{R}^n \to \mathbb{R}^n$ be continuous.

(a) If $\mathbf{F}$ is strictly monotone on $\mathcal{K}$, then VI($\mathcal{K}, \mathbf{F}$) has at most one solution.
(b) If $\mathbf{F}$ is ξ-monotone on $\mathcal{K}$ for some $\xi > 1$, then VI($\mathcal{K}, \mathbf{F}$) has a unique solution.

Proof: Proof of this theorem can be found in [94], Chapter 2, Theorem 2.3.3. Essential outline of the proof is as follows.

(a) Assume that F is strictly monotone on K. If $x \neq x'$ are two distinct solutions of the VI(K, F), $\forall y \in K$, we have

$$(y - x)^T F(x) \geq 0 \quad \text{and} \quad (y - x')^T F(x') \geq 0 \tag{B.6}$$

Substitute $y = x'$ into the first inequality and $y = x$ into the second inequality:

$$(x' - x)^T F(x) \geq 0 \quad \text{and} \quad (x - x')^T F(x') \geq 0 \tag{B.7}$$

Add these two inequalities:

$$(x' - x)^T (F(x') - F(x)) \leq 0 \tag{B.8}$$

This inequality contradicts the strict monotonicity property of F, thus establishing statement (a).

(b) Let $\overline{F} : \mathbb{R}^n \to \mathbb{R}^n$ denote a continuous extension of F, then SOL(K, F) = SOL($K, \overline{F}$). If F is ξ-monotone on K for some $\xi > 1$, then $\exists x^{ref} \in K$ such that the set

$$L_< = \{x \in K |\ F(x)^T(x - x^{ref}) < 0\} \tag{B.9}$$

is bounded (possibly empty). This implies that there exists a bounded open set Ω and a vector $x^{ref} \in K \cap \Omega$ such that

$$F(x)^T(x - x^{ref}) \geq 0,\ \forall x \in K \cap \partial\Omega \tag{B.10}$$

where $\partial\Omega$ denotes the topological boundary of Ω. This implies that the VI(K, F) has a solution. Moreover, if the set

$$L_\leq = \{x \in K |\ F(x)^T(x - x^{ref}) \leq 0\} \tag{B.11}$$

which is nonempty and larger than $L_<$, is bounded, then SOL(K, F) is nonempty and compact. The uniqueness of the solution follows from part (a).

■

Theorem (Equilibrium Points): Assume that $\mathscr{K}$ is a convex polyhedron. Then, the equilibrium points of the PDS$(\mathscr{K}, F)$ coincide with the solutions of VI$(\mathscr{K}, F)$ [221].

Proof: Proof of this theorem can be found in [221], Chapter 2, Theorem 2.4. Essential outline of the proof is as follows.

$$\Pi_K(\mathbf{p}^*, -\mathbf{F}(\mathbf{p}^*)) = 0 \Longleftrightarrow \begin{cases} \text{either } \ \mathbf{F}(\mathbf{p}^*) = 0, \ \text{or} \\ \mathbf{p}^* \in \partial K; \ \ \mathbf{F}(\mathbf{p}^*) = \alpha\mathbf{s}, \ \alpha > 0, \ \mathbf{s} \in S(\mathbf{p}^*) \end{cases} \tag{B.12}$$

which is equivalent to VI(K, F). ■

Theorem (Trajectory Uniqueness): If $\mathbf{F}$ in the initial value problem

$$\dot{\mathbf{p}} = \Pi_{\mathscr{K}}(\mathbf{p}, -\mathbf{F}(\mathbf{p})) \tag{B.13}$$

with

$$\mathbf{p}(t_0) = \mathbf{p}_0 \in \mathscr{K} \tag{B.14}$$

is *Lipschitz* continuous, then for any $\mathbf{p}_0 \in K$, there exists a unique solution $\mathbf{p}(t)$ to the above initial value problem.

Proof: Proof of this theorem for solutions in Euclidean space can be found in [221], Chapter 2, Theorem 2.5. The proof is based on the assumption that F is Lipschitz continuous with linear growth. In [281], Chapter 6, Theorem 6.1 and [282], Theorem 3.1, results were generalized from Euclidean space to Hilbert spaces of arbitrary dimensions. Also, the linear growth condition was relaxed. Essential outline of the proof for

Hilbert spaces is presented later. Essential outline of the proof for Euclidean space is as follows. ■

The associated ODE with discontinuous right-hand side is written as a pair of two equations. The first one is the ODE without the projection operator and the second one is a mapping that restricts the solution of the first equation to K. This approach benefits from the results of the *Skorokhod problem* [283] for finding such a mapping. The Skorokhod problem defines a mapping from the space of paths to itself [221].

Definition (Skorokhod Problem): Let $\psi \in D([0,\infty),\mathbb{R}^{m\times n})$ with $\psi(0) \in K$ be given. Then (ϕ,η) solves the Skorokhod problem with respect to K if $\forall t \in [0,\infty)$

(i) $\phi(t) = \psi(t) + \eta(t),\ \phi(0) = \psi(0)$
(ii) $\phi(t) \in K$
(iii) $|\eta(t)| < \infty$
(iv) $|\eta(t)| = \int_{(0,t]} I_{\partial K}(\phi(\tau))\mathrm{d}|\eta(\tau)|$, where I is an indicator function.
(v) There exists measurable $\gamma : [0,\infty) \to \mathbb{R}^{m\times n}$ such that $\gamma(\tau) \in s(\phi(\tau))$ and $\eta(t) = \int_{(0,t]} \gamma(\tau)\mathrm{d}|\eta(\tau)|$, where s is the inward normal.

Theorem (PDS Equilibrium Characteristics): Suppose that $\mathbf{x}^*$ solves VI($\mathcal{K}$, $\mathbf{F}$).

(a) If the mapping $\mathbf{F}$ is strictly monotone at $\mathbf{x}^*$, then $\mathbf{x}^*$ is a strict monotone attractor for the PDS($\mathcal{K}$, $\mathbf{F}$).
(b) If the mapping $\mathbf{F}$ is ξ-monotone at $\mathbf{x}^*$ with $\xi < 2$, then $\mathbf{x}^*$ is a finite-time attractor.
(c) If the mapping $\mathbf{F}$ is strongly monotone at $\mathbf{x}^*$, then $\mathbf{x}^*$ is exponentially stable.

Proof:

(a) Proof of this theorem can be found in [221], Chapter 3, Theorem 3.6. Essential outline of the proof is as follows.

Consider the Lyapunov function

$$V(t) = \frac{1}{2}\|\mathbf{p}(t) - \mathbf{p}^*\|^2 \tag{B.15}$$

Then

$$\dot{V}(t) = \langle(\mathbf{p}(t) - \mathbf{p}^*), \Pi_K(\mathbf{p}(t), \sigma + \mathbf{M}\mathbf{p}(t))\rangle \tag{B.16}$$

Regarding (4.15), it can be shown that

$$\dot{V}(t) \leq \langle(\mathbf{p}(t) - \mathbf{p}^*), -(\sigma + \mathbf{M}\mathbf{p}(t))\rangle \tag{B.17}$$

Due to strict monotonicity, we have

$$\dot{V}(t) < 0 \tag{B.18}$$

when $\mathbf{p}(t) \neq \mathbf{p}^*$, and

$$\lim_{t\to\infty} V(t) = 0 \tag{B.19}$$

Therefore, $\mathbf{p}^*$ is a is a strict monotone attractor.

(b) Proof of this theorem can be found in [221], Chapter 3, Theorem 3.8. Essential outline of the proof is as follows.

Consider the Lyapunov function $V(t)$ as (B.15). Since ξ-monotonicity implies strict monotonicity, $V(t)$ is strictly decreasing. It can be shown that due to ξ-monotonicity, $V(t)$ reaches zero and then it remains zero. Hence, there exists a T such that

$$\begin{cases} V(t) > 0, t \leq T \\ V(t) = 0, t > T \end{cases} \tag{B.20}$$

Therefore, $\mathbf{p}^*$ is a is a finite-time attractor.

(c) Proof of this theorem can be found in [221], Chapter 3, Theorem 3.7. Essential outline of the proof is as follows.

Consider the Lyapunov function $V(t)$ as (B.15). Regarding (4.15), it can be shown that

$$\dot{V}(t) \leq -||\mathbf{M}||.||\mathbf{p}(t) - \mathbf{p}^*||^2 \tag{B.21}$$

If there exists a $t_0 \geq 0$ for which $||\mathbf{p}(t_0) - \mathbf{p}^*|| = 0$, we have

$$||\mathbf{p}(t) - \mathbf{p}^*|| = 0, \ \forall t \geq t_0 \tag{B.22}$$

Since strong monotonicity implies monotonicity, we have

$$||\mathbf{p}(t) - \mathbf{p}^*|| \leq ||\mathbf{p}_0 - \mathbf{p}^*|| \leq c||\mathbf{p}_0 - \mathbf{p}^*||e^{-\eta t} \tag{B.23}$$

where $c = e^{\eta t_0}$. Assume that

$$||\mathbf{p}(t) - \mathbf{p}^*|| \neq 0, \ \forall t \geq 0 \tag{B.24}$$

Dividing both sides of (B.21) by $V(t)$ and taking the integral, we obtain

$$||\mathbf{p}(t) - \mathbf{p}^*|| \leq c'||\mathbf{p}_0 - \mathbf{p}^*||e^{-\eta t} \tag{B.25}$$

Therefore, $\mathbf{p}^*$ is exponentially stable. ■

Theorem (Robust Exponential Stability): Consider the system (4.34) with initial condition (4.30), and assume that $-\mathbf{M}$ is a Hurwitz stable matrix satisfying

$$\|e^{\mathbf{M}t}\| \leq ce^{-\eta t} \tag{B.26}$$

for some real numbers $c \geq 1$ and $\eta > 0$. In the left-hand side of the above equation, e denotes a "matrix" exponential operator. If the inequality

$$\frac{c}{\eta}\left[\overline{\tau}\sum_{\ell=1}^{m(m-1)}(\mu_1^\ell + \mu_2^\ell) + b_d + \sum_{\ell=1}^{m(m-1)} b_d^\ell\right] < 1 \tag{B.27}$$

holds, then the transient response of $\mathbf{p}(t)$ satisfies

$$\|\mathbf{p}(t)\| \leq \zeta \sup_{\theta\in\Psi_{t_0}} \{\|\boldsymbol{\phi}(\theta)\|\}e^{-\rho\int_{t_0}^{t}\frac{d\theta}{\tau(\theta)}}, \ \forall t \geq t_0, \zeta \geq 1 \tag{B.28}$$

where

$$\mu_1^\ell = \|\mathbf{M}_d^\ell\| + \|\mathbf{M}_d^\ell\|b_d \tag{B.29}$$

$$\mu_2^\ell = \sum_{j=1}^{m(m-1)} \|\mathbf{M}_d^\ell\mathbf{M}_d^j\| + \|\mathbf{M}_d^\ell\| \sum_{j=1}^{m(m-1)} b_d^j \tag{B.30}$$

and $\rho > 0$ is the unique positive solution of the transcendental equation

$$1 - \frac{c}{\eta}b_d - \frac{\rho}{\eta\tau(0)} = \mu_3\frac{c}{\eta}e^{\frac{\rho}{1-\delta}} \tag{B.31}$$

where

$$\mu_3 = \overline{\tau}\sum_{\ell=1}^{m(m-1)}\mu_1^\ell + \overline{\tau}e^{\frac{\rho}{1-\delta}}\sum_{\ell=1}^{m(m-1)}\mu_2^\ell + \sum_{\ell=1}^{m(m-1)} b_d^\ell \tag{B.32}$$

Furthermore, the system described by (4.34) and (4.30) is robustly exponentially stable with a decay rate $\rho/\overline{\tau}$.

Proof: The proof uses ideas given in [232, 284]. Let us consider the following differential equation:

$$\dot{\mathbf{y}}(t) = -(\eta - cb_d)\mathbf{y}(t) + q(t)\mathbf{y}(t - \tau(t)) \tag{B.33}$$

where

$$q(t) = \left(\eta - cb_d - \frac{\rho}{\tau(t)}\right)e^{-\rho\int_{t-\tau(t)}^{t}\frac{d\theta}{\tau(\theta)}} \tag{B.34}$$

It can be verified that

$$\mathbf{y}(t) = C_0 e^{-\rho \int_{t_0}^{t} \frac{d\theta}{\tau(\theta)}} \tag{B.35}$$

is a solution of (B.33), where C_0 is a constant. The mean-value theorem is applied to $\int_{t-\tau(t)}^{t} \frac{d\theta}{\tau(\theta)}$ twice . It follows that $\exists\ \theta_1, \theta_2 \in \mathbb{R}$ satisfying $0 < \theta_1 < \theta_2 < 1$ such that

$$\begin{aligned}\int_{t-\tau(t)}^{t} \frac{d\theta}{\tau(\theta)} &= \frac{\tau(t)}{\tau(t) - \theta_1 \tau(t)\dot{\tau}(t - \theta_2 \tau(t))} \\ &= \frac{1}{1 - \theta_1 \dot{\tau}(t - \theta_2 \tau(t))} \leq \frac{1}{1-\delta}\end{aligned} \tag{B.36}$$

For $\rho > 0$ satisfying (B.31), we have

$$q(t) \geq \left(\eta - cb_d - \frac{\rho}{\tau(t_0)}\right) e^{-\frac{\rho}{1-\delta}} = c\mu_3 \tag{B.37}$$

Now we show that for a proper choice of C_0, the solution of (B.35) is an upper bound for the solution of (4.34) and (4.30).

Let us choose C_0 such that the following inequalities are satisfied simultaneously:

$$\mathbf{y}(t) \geq \|\boldsymbol{\phi}(\theta)\|,\ \forall \theta \in \psi_{t_0} \tag{B.38}$$

$$C_0 \geq c \sup_{\theta \in \Psi_{t_0}} \|\boldsymbol{\phi}(\theta)\| \tag{B.39}$$

Solution of (4.34) can be written as

$$\begin{aligned}\mathbf{p}(t) = \mathbf{p}(t_0)e^{-\mathbf{I}t} &- \int_{t_0}^{t} e^{-\mathbf{I}(t-\theta)} \sum_{\ell=1}^{m(m-1)} \mathbf{M}_d^{\ell} \mathbf{p}(\theta - \tau^{\ell}(\theta))d\theta \\ &- \int_{t_0}^{t} e^{-\mathbf{I}(t-\theta)}[\varrho(\theta) + \Delta\mathbf{M}_d^{\ell}(\theta)\mathbf{p}(\theta - \tau^{\ell}(\theta))]d\theta\end{aligned} \tag{B.40}$$

Regarding (4.37), (4.38), and (B.26), $\forall t \geq t_0$ we have

$$\begin{aligned}\|\mathbf{p}(t)\| \leq ce^{-\eta t} \sup_{\theta \in \Psi_{t_0}} \|\boldsymbol{\phi}(\theta)\| &+ \int_{t_0}^{t} ce^{-\eta(t-\theta)} b_d \|\mathbf{p}(\theta)\| d\theta \\ &+ \int_{t_0}^{t} ce^{-\eta(t-\theta)} \sum_{\ell=1}^{m(m-1)} (\|\mathbf{M}_d^{\ell}\| + b_d^{\ell})\|\mathbf{p}(\theta - \tau^{\ell}(\theta))\| d\theta\end{aligned} \tag{B.41}$$

Considering the term $cb_d\mathbf{y}(t) + q(t)\mathbf{y}(t - \tau(t))$ in (B.33) as an inhomogeneous term, The solution of this equation can be written as

$$\mathbf{y}(t) = C_0 e^{-\eta t} + \int_{t_0}^{t} cb_d e^{-\eta(t-\theta)} \mathbf{y}(\theta)d\theta + \int_{t_0}^{t} e^{-\eta(t-\theta)} q(\theta)\mathbf{y}(\theta - \tau(\theta))d\theta \tag{B.42}$$

In order to compare $\|\mathbf{p}(t)\|$ with $\mathbf{y}(t)$, we define $\mathbf{z}(t) = \|\mathbf{p}(t)\| - \mathbf{y}(t)$. From (B.41) and (B.42), $\forall t \geq t_0$ we have

$$\mathbf{z}(t) \leq \left(c \sup_{\theta \in \Psi_{t_0}} \|\boldsymbol{\phi}(\theta)\| - C_0 \right) e^{-\eta t} + c \int_{t_0}^{t} e^{-\eta(t-\theta)} \left(\sum_{\ell=1}^{m(m-1)} (\|\mathbf{M}_d^{\ell}\| + b_d^{\ell}) \mathbf{z}(\theta - \tau(\theta)) + b_d \mathbf{z}(\theta) \right) \mathrm{d}\theta + \int_{t_0}^{t} e^{-\eta(t-\theta)} \left(c \sum_{\ell=1}^{m(m-1)} (\|\mathbf{M}_d^{\ell}\| + b_d^{\ell}) - q(\theta) \right) \mathbf{y}(\theta - \tau(\theta)) \mathrm{d}\theta \tag{B.43}$$

Inequalities (B.37) and (B.38) lead to

$$\mathbf{z}(t) \leq c \int_{t_0}^{t} e^{-\eta(t-\theta)} \left(\sum_{\ell=1}^{m(m-1)} (\|\mathbf{M}_d^{\ell}\| + b_d^{\ell}) \mathbf{z}(\theta - \tau(\theta)) + b_d \mathbf{z}(\theta) \right) \mathrm{d}\theta \tag{B.44}$$

Also, (B.38) implies that

$$\mathbf{z}(t) \leq 0, \ \forall t \in \Psi_{t_0} \tag{B.45}$$

Since $\mathbf{z}(t)$ is continuous, the above inequality holds in some neighborhood of t_0. Assume that t^* is the smallest t for which $\mathbf{z}(t^*) > 0$. Due to (B.45) and the fact that $\mathbf{z}(\theta) \leq 0$ for any $0 \leq \theta \leq t^*$, it follows from (B.44) that $\mathbf{z}(t^*) \leq 0$, which contradicts the assumption. Hence, $\mathbf{z}(t) \leq 0$ for all $t \geq t_0$, which leads to

$$\|\mathbf{p}(t)\| \leq C_0 e^{-\rho \int_{t_0}^{t} \frac{\mathrm{d}\theta}{\tau(\theta)}} = \zeta \sup_{\theta \in \Psi_{t_0}} \{\|\boldsymbol{\phi}(\theta)\|\} e^{-\rho \int_{t_0}^{t} \frac{\mathrm{d}\theta}{\tau(\theta)}}, \ \forall t \geq t_0 \tag{B.46}$$

where

$$\zeta = \frac{C_0}{\sup_{\theta \in \Psi_{t_0}} \|\boldsymbol{\phi}(\theta)\|} \geq 1 \tag{B.47}$$

Finally, from the boundedness of $\tau(t)$, we have

$$\|\mathbf{p}(t)\| \leq \zeta \sup_{\theta \in \Psi_{t_0}} \{\|\boldsymbol{\phi}(\theta)\|\} e^{-\frac{\rho}{\tau}}, \ \forall t \geq t_0 \tag{B.48}$$

Therefore, the system (4.34) and (4.30) is robustly exponentially stable with a decay rate $\frac{\rho}{\tau}$. ■

Theorem (Trajectory Uniqueness in Hilbert Space): Let H be a Hilbert space and K be a nonempty, closed, convex subset. Let $F : \mathscr{K} \rightarrow H$ be a Lipschitz-continuous vector field and $\mathbf{p}_0 \in \mathscr{K}$. Then the initial value problem

$$\frac{d\mathbf{p}(\tau)}{d\tau} = \Pi_{\mathscr{K}}(\mathbf{p}(\tau), -\mathbf{F}(\mathbf{p}(\tau))), \;\; \mathbf{p}(0) = \mathbf{p}_0 \in \mathscr{K} \tag{B.49}$$

has a unique absolutely continuous solution on the interval $[0, \infty)$.

Proof: Proof of this theorem can be found in [281], Chapter 6, Theorem 6.1 and [282], Theorem 3.1. Essential outline of the proof is as follows.

Let L be the Lipschitz constant and $\|\mathbf{p}\| \leq b$ for $b > 0$. Consider the interval $[0, l]$ where $l = \frac{b}{\|\mathbf{F}(\mathbf{p}_0)\| + bL}$.

(i) Construct the sequence $\{\mathbf{p}_k(.)\}$ of absolutely continuous functions defined on $[0, l]$ with values in $\mathscr{K}$ such that $\forall k \geq 0 \;\; \mathbf{p}_k(0) = \mathbf{p}_0$ and for almost all $t \in [0, l]$ and every neighborhood $\mathscr{M} \in \mathscr{K} \times \mathscr{K}$ of 0, $\{\mathbf{p}_k(.)\}$ and the sequence of its derivatives $\{\mathbf{p}'_k(.)\}$ have the following property:

$$(\mathbf{p}_k(t), \mathbf{p}'_k(t)) \in \text{graph}(\mathbf{F} - \tilde{N}_{\mathscr{K}}) + \mathscr{M}, \;\; \forall k \geq k_0(t, \mathscr{M}) \tag{B.50}$$

where

$$\tilde{N}_{\mathscr{K}} k(\mathbf{p}) = \{\mathbf{n} \in N_{\mathscr{K}} k(\mathbf{p}) | \; \|\mathbf{n}\| \leq \|\mathbf{F}(\mathbf{p})\|\} \subseteq N_{\mathscr{K}}(\mathbf{p}) \tag{B.51}$$

and $N_{\mathscr{K}}(\mathbf{p})$ is the normal cone to the set $\mathscr{K}$ at the point $\mathbf{p}$:

$$N_{\mathscr{K}}(\mathbf{p}) = \{\mathbf{n} \in \mathscr{K} \,|\, \langle \mathbf{n}, \mathbf{p} - \mathbf{x} \rangle \geq 0, \;\; \forall \mathbf{x} \in \mathscr{K}\} \tag{B.52}$$

The uniform convergence of the sequence $\{\mathbf{p}_k(.)\}$ can be proved.

(ii) After proving the uniform convergence of the sequence of approximate solutions $\{\mathbf{p}_k(.)\}$ to a limit $\mathbf{p}(.)$, select a subsequence for which the sequence of derivatives $\{\mathbf{p}'_k(.)\}$ in $L^{\infty}([0, l], \mathscr{K})$ converges weakly to the derivative of $\mathbf{p}(.)$.

(iii) It is shown that $\mathbf{p}(.)$ is a solution to the initial value problem (B.49).

(iv) From *(i)*–*(iii)*, we know that the problem has solutions on the interval $[0, l]$. Assume that $\mathbf{p}_1(.)$ and $\mathbf{p}_2(.)$ are two solutions starting at the point $\mathbf{p}_0$. It is shown that

$$\|\mathbf{p}_1(t) - \mathbf{p}_2(t)\|^2 \leq 0 \tag{B.53}$$

Therefore, $\mathbf{p}_1(t) = \mathbf{p}_2(t), \;\; \forall t \in [0, l]$, which proves the uniqueness of the solutions.

(v) Having the unique solution for the interval $[0, l]$, we consider $t_0 = l$ and apply the theorem again. Hence, we obtain a solution for an extended time interval. By continuing this process, a solution can be obtained for $t \in [0, \infty)$. ■

Theorem (Existence and Uniqueness of EVI/PDS$_t$ Equilibria): If $\mathbf{F}(\mathbf{p}) = \sigma + \mathbf{M}\mathbf{p}$ is strictly monotone and Lipschitz continuous on $\mathcal{K}$, then there exists $\mathbf{p}^* \in K$ such that

- $\mathbf{p}^*$ uniquely solves the EVI problem.
- $\mathbf{p}^*$ uniquely solves $\Pi_{\mathcal{K}}(\mathbf{p}(.,\tau), -\mathbf{F}(\mathbf{p}(.,\tau))) = 0$.

Proof: The proof uses ideas given in [94, 223, 285].

(a) Uniqueness of the solution is concluded from Theorem (Existence and Uniqueness of VI Solution), and the fact that F is an affine mapping.

(b) In [223], Proposition 3.1, it is proved that the PDS has at most one equilibrium point. Essential outline of the proof is as follows.

Assume that the PDS has at least two solutions $\mathbf{p}_1 \neq \mathbf{p}_2 \in K$. Then,

$$\Pi_K(\mathbf{p}_1, -\mathbf{F}(\mathbf{p}_1)) = 0 \text{ and } \Pi_K(\mathbf{p}_2, -\mathbf{F}(\mathbf{p}_2)) = 0 \tag{B.54}$$

Equivalently, this means that $-\mathbf{F}(\mathbf{p}_1) \in N_K(\mathbf{p}_1)$ and $-\mathbf{F}(\mathbf{p}_2) \in N_K(\mathbf{p}_2)$, where N_K is the normal cone. Since the set-valued mapping $\mathbf{p} \to N_K(\mathbf{p})$ is a monotone mapping, we have

$$(-\mathbf{F}(\mathbf{p}_1) + \mathbf{F}(\mathbf{p}_2))^T(\mathbf{p}_1 - \mathbf{p}_2) \geq 0 \tag{B.55}$$

or equivalently

$$(\mathbf{F}(\mathbf{p}_2) - \mathbf{F}(\mathbf{p}_1))^T(\mathbf{p}_2 - \mathbf{p}_1) \leq 0 \tag{B.56}$$

On the other hand, from strict monotonicity property of F, we have

$$(\mathbf{F}(\mathbf{p}_2) - \mathbf{F}(\mathbf{p}_1))^T(\mathbf{p}_2 - \mathbf{p}_1) > 0 \tag{B.57}$$

The last two equations lead to a contradiction. Therefore, the PDS has at most one equilibrium point. Solutions of the EVI problem are the same as the stationary points of the PDS and vice versa [223]. From (a) we know that the EVI has a unique solution. Therefore, the PDS has a unique equilibrium as well. ∎

Theorem (PDS$_t$ Equilibria Characteristics): Assume $\mathbf{F} : \mathcal{K} \to \ell_2([0, T], \mathbb{R}^{m \times n})$ is Lipschitz continuous on $\mathcal{K}$

- If $\mathbf{F}$ is strictly pseudomonotone on $\mathcal{K}$, then the unique curve of equilibria is a strict monotone attractor.
- If $\mathbf{F}$ is strongly pseudomonotone on $\mathcal{K}$, then the unique curve of equilibria is exponentially stable and an attractor.

Proof: Proof of this theorem can be found in [281], Chapter 7, Theorem 7.2, and Theorem 7.6. Essential outline of the proof is as follows.

(a) Consider the Lyapunov function

$$V(t) = \frac{1}{2}\|\mathbf{p}(t) - \mathbf{p}^*\|^2 \tag{B.58}$$

Then

$$\dot{V}(t) = \langle(\mathbf{p}(t) - \mathbf{p}^*), \Pi_{\mathcal{K}}(\mathbf{p}(t), -(\sigma + \mathbf{M}\mathbf{p}(t)))\rangle \tag{B.59}$$

Regarding (4.15), we have

$$\dot{V}(t) \leq \langle(\mathbf{p}(t) - \mathbf{p}^*), -(\sigma + \mathbf{M}\mathbf{p}(t))\rangle \tag{B.60}$$

Strict pseudomonotonicity leads to

$$\langle(\mathbf{p}(t) - \mathbf{p}^*), -(\sigma + \mathbf{M}\mathbf{p}(t))\rangle < 0 \tag{B.61}$$

and

$$\dot{V}(t) < 0 \tag{B.62}$$

Therefore, $\mathbf{p}^*$ is a is a strict monotone attractor.

(b) Since F is strongly pseudomonotone, there exists $c > 0$ such that

$$\langle\mathbf{F}(\mathbf{p}^*), \mathbf{p} - \mathbf{p}^*\rangle \geq 0 \Longrightarrow \langle\mathbf{F}(\mathbf{p}), \mathbf{p} - \mathbf{p}^*\rangle \geq c\|\mathbf{p} - \mathbf{p}^*\|^2, \ \forall \mathbf{p} \in \mathcal{K} \tag{B.63}$$

which implies that

$$\langle\mathbf{p}(t) - \mathbf{p}^*, -(\sigma + \mathbf{M}\mathbf{p}(t))\rangle \leq -c\|\mathbf{p} - \mathbf{p}^*\|^2 \tag{B.64}$$

From (B.59), we have

$$\dot{V}(t) \leq -c\|\mathbf{p} - \mathbf{p}^*\|^2 \tag{B.65}$$

Integration of the above inequality leads to

$$\frac{1}{2}\|\mathbf{p}(t) - \mathbf{p}^*\|^2 \leq \|\mathbf{p}_0 - \mathbf{p}^*\|^2 e^{-ct} \Rightarrow \|\mathbf{p}(t) - \mathbf{p}^*\| \leq \sqrt{2}\|\mathbf{p}_0 - \mathbf{p}^*\| e^{-\frac{c}{2}t} \tag{B.66}$$

which shows that $\mathbf{p}^*$ is exponentially stable. As $t \to \infty$, we obtain $\mathbf{p}(t) \to \mathbf{p}^*$ and, therefore, $\mathbf{p}^*$ is an attractor. ∎

REFERENCES

1. M. Hermann, T. Pentek, and B. Otto, "Design principles for industrie 4.0 scenarios: A literature review," *Working Paper 01/2015*, Technische Universität Dortmund, pp. 1–15, 2015.
2. D. Boswarthick, O. Elloumi, and O. Hersent, *M2M Communications: A Systems Approach*. John Wiley and Sons, 2012.
3. J. Höller, V. Tsiatsis, C. Mulligan, S. Karnouskos, and D. Boyle, *From Machine-to-Machine to the Internet of Things: Introduction to a New Age of Intelligence*. Academic Press, 2014.
4. V. B. Mišić and J. Mišić, *Machine-to-Machine Communications: Architectures, Technology, Standards, and Applications*. CRC Press, 2014.
5. R. Stackowiak, A. Licht, V. Mantha, and L. Nagode, *Big Data and the Internet of Things: Enterprise Information Architecture for A New Age*. APress, 2015.
6. C. Anton-Haro and M. Dohler, *Machine-to-Machine (M2M) Communications: Architecture, Performance and Applications*. Woodhead Publishing, 2015.
7. D. B. Rawat, J. J. P. C. Rodrigues, and I. Stojmenovic, *Cyber-Physical Systems: From Theory to Practice*. CRC Press, 2015.
8. R. Poovendran, K. Sampigethaya, S. K. S. Gupta, I. Lee, K. V. Prasad, D. Corman, and J. Paunicka, "Special issue on cyber – physical systems [Scanning the Issue]," *Proceedings of the IEEE*, vol. 100, no. 1, pp. 6–12, 2012.
9. S. Soatto, "Actionable information in vision," in *Machine Learning for Computer Vision*, R. Cipolla, S. Battiato, and G. M. Farinella, Eds., Springer-Verlag, pp. 17–48, 2013.
10. S. Haykin, M. Fatemi, P. Setoodeh, and Y. Xue, "Cognitive control," *Proceedings of the IEEE*, vol. 100, no. 12, pp. 3156–3169, 2012.
11. M. E. Porter, *Competitive Strategy: Techniques for Analyzing Industries and Competitors*. Free Press, 1980.
12. H. Kagermann, W. Wahlster, and J. Helbig, "Recommendations for implementing the strategic initiative Industrie 4.0," *Final report of the Industrie 4.0 Working Group*, pp. 1–82, 2013.

Fundamentals of Cognitive Radio, First Edition. Peyman Setoodeh and Simon Haykin.

13. S. Haykin, *Cognitive Dynamic Systems*. Cambridge University Press, 2012.
14. A. J. I. Jones, A. Artikis, and J. Pitt, "The design of intelligent socio-technical systems," *Artificial Intelligence Review*, vol. 39, no. 1, pp. 5–20, 2013.
15. J. Pitt, J. Schaumeier, and A. Artikis, "Axiomatization of socio-economic principles for self-organizing institutions: Concepts, experiments, and challenges," *ACM Transactions on Autonomous and Adaptive Systems*, vol. 7, no. 4, pp. 39:1–39:39, 2012.
16. E. Ostrom, *Governing the Commons*. Cambridge University Press, 1990.
17. S. Chopra and P. Meindl, *Supply Chain Management: Strategy, Planning, and Operation*. Prentice Hall, 5th ed., 2012.
18. A. Nagurney, *Supply Chain Network Economics: Dynamics of Prices, Flows, and Profits*. Edward Elgar Publishing, 2006.
19. A. Goldsmith, *Wireless Communications*. Cambridge University Press, 2005.
20. P. J. Nahin, "Maxwell's grand unification," *IEEE Spectrum*, vol. 29, no. 3, p. 45, 1992.
21. J. C. Maxwell, "A dynamical theory of the electromagnetic field," *Philosophical Transactions of the Royal Society of London*, vol. 155, pp. 459–512, 1865.
22. J. C. Maxwell, *A Treatise on Electricity and Magnetism*. Volumes I and II, MacMillan and Co., 1873.
23. P. J. Nahin, *Oliver Heaviside: Sage in Solitude*. IEEE Press, 1988.
24. H. Hertz, *Electric Waves: Being Researches on the Propagation of Electric Action with Finite Velocity Through Space*. MacMillan and Co., 1893.
25. C. Susskind, *Heinrich Hertz*. San Francisco Press Inc., 1995.
26. P. Stuckmann and R. Zimmermann, "Toward ubiquitous and unlimited-capacity communication networks: European research in framework programme 7," *IEEE Communications Magazine*, vol. 45, no. 5, pp. 148–157, 2007.
27. L. Doyle, J. Kibilda, T. K. Forde, and L. Dasilva, "Spectrum without bounds, networks without borders," *Proceedings of the IEEE*, vol. 102, no. 3, pp. 351–365, 2014.
28. W. W. Lu, "Point-of-view article: Opening the U.S. mobile communications market," *Proceedings of the IEEE*, vol. 96, no. 9, pp. 1450–1452, 2008.
29. J. Mitola, *Cognitive Radio: An Integrated Agent Architecture for Software Defined Radio*. Ph.D. Dissertation. Royal Institute Technolology (KTH), Stockholm, Sweden, 2000.
30. S. Haykin, "Cognitive radio: Brain-empowered wireless communications," *IEEE Journal on Selected Areas in Communications*, vol. 23, no. 2, pp. 201–220, 2005.
31. E. Biglieri, A. J. Goldsmith, L. J. Greenstein, N. B. Mandayam, and H. V. Poor, *Principles of Cognitive Radio*. Cambridge University Press, 2013.
32. A. J. Fehske, I. Viering, J. Voigt, and C. Sartori, "Small-cell self-organizing wireless networks," *Proceedings of the IEEE*, vol. 102, no. 3, pp. 334–350, 2014.
33. G. Gür, S. Bayhan, and F. Alagoz, "Cognitive femtocell networks: An overlay architecture for localized dynamic spectrum access," *IEEE Wireless Communications*, vol. 17, no. 4, pp. 62–70, 2010.
34. S. Haykin, "Radar vision," in *Proceedings of the International Specialist Seminar on the Design and Application of Parallel Digital Processors*, 1991, pp. 75–78.
35. J. Mitola and G. Q. Maguire Jr., "Cognitive radio: Making software radios more personal," *IEEE Personal Communications*, vol. 6, no. 4, pp. 13–18, 1999.
36. J. M. Fuster, *Cortex and Mind: Unifying Cognition*. Oxford University Press, 2003.

37. S. Haykin and P. Setoodeh, "Cognitive radio networks: The spectrum supply chain paradigm," *IEEE Transactions on Cognitive Communications and Networking*, vol. 1, no. 1, pp. 3–28, 2015.
38. FCC, "*Spectrum policy task force report*," ET Docket No. 02-135, 2002.
39. M. McHenry, "Report on spectrum occupancy measurements," *Shared Spectrum Company*, http://www.sharedspectrum.com/, 2005.
40. M. A. McHenry, P. A. Tenhula, D. McCloskey, D. A. Roberson, and C. S. Hood, "Chicago spectrum occupancy measurements and analysis and a long-term studies proposal," in *Proceedings of Workshop on Technology and Policy for Accessing Spectrum (TAPAS)*, 2006, pp. 1–12.
41. T. Erpek, M. Lofquist, and K. Patton, "Spectrum occupancy measurements: Loring Commerce Centre, Limestone, Maine, Sep. 18-20, 2007," *Technical Report, Shared Spectrum Company*, pp. 1–35, 2007.
42. A. Petrin and P. G. Steffes, "Analysis and comparison of spectrum measurements performed in urban and rural areas to determine the total amount of spectrum usages," in *Proceedings of the International Symposium on Advanced Radio Technologies*, 2005, pp. 9–12.
43. R. I. Chiang, G. B. Rowe, and K. W. Sowerby, "A quantitative analysis of spectral occupancy measurements for cognitive radio," in *Proceedings of IEEE Vehicular Technology Conference*, 2007, pp. 3016–3020.
44. M. H. Islam, C. L. Koh, S. W. Oh, X. Qing, Y. Y. Lai, C. Wang, Y. C. Liang, B. E. Toh, F. Chin, G. L. Tan, and W. Toh, "Spectrum survey in Singapore: occupancy measurements and analysis," in *Proceedings of the 3rd International Conference of Cognitive Radio Oriented Wireless Network and Communications (CROWNCOM08)*, 2008, pp. 1–7.
45. K. A. Qaraqe, H. Celebi, M. S. Alouini, A. El-Saigh, L. Abuhantash, M. Al-Mulla, O. Al-Mulla, A. Jolo, and A. Ahmed, "Measurement and analysis of wideband spectrum utilization in indoor and outdoor environments," in *Proceedings of the IEEE 18th Signal Processing and Communications Applications Conference (SIU-2010)*, 2010, pp. 1–4.
46. F. Khozeimeh and S. Haykin, "Dynamic spectrum management for cognitive radio: An overview," *Wireless Communications and Mobile Computing*, vol. 9, no. 11, pp. 1447–1459, 2009.
47. F. Khozeimeh, *Self-organizing Dynamic Spectrum Management: Novel Scheme for Cognitive Radio Networks*. Ph.D. Dissertation. McMaster University, Canada, 2011.
48. S. Haykin, *Neural Networks and Learning Machines*. Prentice Hall, 3rd ed., 2009.
49. M. D. Mesarovic and Y. Takahara, *Abstract Systems Theory*. Springer-Verlag, 1989.
50. B. Wang, Y. Wu, and K. J. R. Liu, "Game theory for cognitive radio networks: An overview," *Computer Networks*, vol. 54, no. 14, pp. 2537–2561, 2010.
51. Y. Tachwali, B. F. Lo, I. F. Akyildiz, and R. Agusti, "Multiuser resource allocation optimization using bandwidth-power product in cognitive radio networks," *IEEE Journal of Selected Areas in Communications*, vol. 31, no. 3, pp. 451–463, 2013.
52. A. Clegg and A. Weisshaar, "Future radio spectrum access," *Proceedings of the IEEE*, vol. 102, no. 3, pp. 239–241, 2014.
53. G. Hattab and M. Ibnkahla, "Multiband spectrum access: great promises for future cognitive radio networks," *Proceedings of the IEEE*, vol. 102, no. 3, pp. 282–306, 2014.

54. M. Li, S. Salinas, P. Li, X. Huang, Y. Fang, and S. Glisic, "Optimal scheduling for multi-radio multi-channel multi-hop cognitive cellular networks," *IEEE Transactions on Mobile Computing*, vol. 14, no. 1, pp. 139–154, 2015.
55. M. J. Marcus, "Harmful interference and its role in spectrum policy," *Proceedings of the IEEE*, vol. 102, no. 3, pp. 265–269, 2014.
56. J. Li, S. Li, F. Zhao, and R. Du, "Co-channel interference modeling in cognitive wireless networks," *IEEE Transactions on Communications*, vol. 62, no. 9, pp. 3114–3128, 2014.
57. Y. Zhang, *Resource Allocation for OFDM-Based Cognitive Radio Systems*. Ph.D. Dissertation. University of British Columbia, Canada, 2008.
58. E. Hossain, D. Niyato, and Z. Han, *Dynamic Spectrum Access and Management in Cognitive Radio Networks*. Cambridge University Press, 2009.
59. P. Setoodeh, S. Haykin, and K. Rezaei-Moghadam, "Dynamic spectrum supply chain model for cognitive radio networks," in *Proceedings of the IEEE International Workshop on Emerging Cognitive Radio Applications and Algorithms (CORAL)*, 2012, pp. 1–6.
60. M. Hugos, *Essentials of Supply Chain Management*. John Wiley and Sons, 3rd ed., 2011.
61. E. Axell, G. Leus, E. G. Larsson, and H. V. Poor, "Spectrum sensing for cognitive radio: State-of-the-art and recent advances," *IEEE Signal Processing Magazine*, vol. 29, no. 3, pp. 101–116, 2012.
62. D. Zhang, R. Shinkuma, and N. Mandayam, "Bandwidth exchange: An energy conserving incentive mechanism for cooperation," *IEEE Transactions on Wireless Communications*, vol. 9, no. 6, pp. 2055–2065, 2010.
63. Y. Xiao, G. Bi, D. Niyato, and L. DaSilva, "A hierarchical game theoretic framework for cognitive radio networks," *IEEE Journal on Selected Areas in Communications*, vol. 30, no. 10, pp. 2053–2069, 2012.
64. J. Y. Halpern, "Beyond Nash equilibrium: Solution concepts for the 21st century," in *Proceedings of the Annual ACM Symposium on Principles of Distributed Computing*, 2008, pp. 1–10.
65. R. Kleinberg, K. Ligett, G. Piliouras, and E. Tardos, "Beyond the Nash equilibrium barrier," in *Proceedings of Symposium on Innovations in Computer Science*, 2011, pp. 125–140.
66. P. Setoodeh and S. Haykin, "Robust transmit power control for cognitive radio," *Proceedings of the IEEE*, vol. 97, no. 5, pp. 915–939, 2009.
67. J. S. Pang, G. Scutari, D. P. Palomar, and F. Facchinei, "Design of cognitive radio systems under temperature-interference constraints: A variational inequality approach," *IEEE Transactions on Signal Processing*, vol. 58, no. 6, pp. 3251–3271, 2010.
68. M. J. Osborne and A. Rubenstein, *A Course in Game Theory*. The MIT Press, 1994.
69. B. B. de Mesquita, *The Predictioneer's Game: Using the Logic of Brazen Self-Interest to See and Shape the Future*. Random House, 2009.
70. J. von Neumann and O. Morgenstern, *Theory of Games and Economic Behavior*. Princeton University Press, 1944.
71. D. D. Siljak, *Decentralized Control of Complex Systems*. Academic Press, 1991.
72. A. G. O. Mutambara, *Decentralized Estimation and Control for Multisensor Systems*. CRC Press, 1998.
73. H. W. Kuhn and S. Nasar, *The Essential John Nash*. Princeton University Press, 2002.

74. J. F. Nash, "Equilibrium points in n-person games," *Proceedings of the National Academy of Sciences of the United States of America*, vol. 36, pp. 48–49, 1950.
75. J. F. Nash, "Non-cooperative games," *Annals of Mathematics*, vol. 54, pp. 286–295, 1951.
76. T. Başar and G. J. Olsder, *Dynamic Noncooperative Game Theory*. SIAM, 1999.
77. T. Başar and P. Bernhard, *H_∞ Optimal Control and Related Minimax Design Problems: A Dynamic Game Approach*. 2nd ed., Birkhauser, 1995.
78. E. Altman, T. Boulogne, R. El-Azouzi, T. Jimenez, and L. Wynter, "A survey on networking games in telecommunication," *Computers and Operations Research*, vol. 24, no. 3, pp. 58–68, 2007.
79. A. MacKenzie and L. DaSilva, *Game Theory for Wireless Engineers*. Morgan and Claypool Publishers, 2006.
80. M. Felegyhazi and J. P. Hubaux, "Game theory in wireless networks: A tutorial," Technical Report LCA-Report-2006-002. 6, EPFL, pp. 1–15, 2007.
81. S. Lasaulce, M. Debbah, and E. Altman, "Methodologies for analyzing equilibria in wireless games," *IEEE Signal Processing Magazine*, vol. 26, no. 5, pp. 41–52, 2009.
82. W. Saad, Z. Han, M. Debbah, A. Hjørungnes, and T. Başar, "Coalitional game theory for communication networks," *IEEE Signal Processing Magazine*, vol. 26, no. 5, pp. 77–97, 2009.
83. F. J. Doyle, "Robust control in biology: From genes to cells to systems," in *Proceedings of the International Federation of Automatic Control (IFAC)*, 2008, pp. 3470–3479.
84. J. B. Burl, *Linear Optimal Control: H_2 and H_∞ Methods*. Addison-Wesley, 1999.
85. V. I. Zhukovskiy and M. E. Salukvadze, *The Vectror-Valued Maximin*. Academic Press, 1994.
86. A. Nagurney and Q. Qiang, *Fragile Networks: Identifying Vulnerabilities and Synergies in an Uncertain World*. John Wiley and Sons, 2009.
87. S. Haykin, D. J. Thomson, and J. H. Reed, "Spectrum sensing for cognitive radio," *Proceedings of the IEEE*, vol. 97, no. 5, pp. 849–877, 2009.
88. M. Fukushima, "Stochastic and robust approaches to optimization problems under uncertainty," in *Proceedings of the International Conference on Informatics Research for Development of Knowledge Society Infrastructure (ICKS)*, 2007, pp. 87–94.
89. S. Forge, R. Horvitz, and C. Blackman, "Perspectives on the value of shared spectrum access," *Final Report for the European Commission*, 2012.
90. S. Ghafoor, P. D. Sutton, C. J. Sreenan, and K. N. Brown, "Cognitive radio for disaster response networks: Survey, potential, and challenges," *IEEE Wireless Communications*, vol. 21, no. 5, pp. 70–80, 2014.
91. N. Uchida, K. Takahata, Y. Shibata, and N. Shiratori, "Evaluation of never die network for a rural area in a ultra large scale disaster," in *Proceedings of the IEEE International Conference on Complex, Intelligent, and Software Intensive Systems*, 2012, pp. 306–313.
92. N. Uchida, N. Kawamura, and Y. Shibata, "Evaluation of cognitive wireless based delay tolerant network for disaster information system in a rural area," in *Proceedings of the IEEE International Conference on Complex, Intelligent, and Software Intensive Systems*, 2013, pp. 1–7.
93. J. M. Danskin, *The Theory of Max-Min*. Springer-Verlag, 1967.
94. F. Facchinei and J. S. Pang, *Finite-Dimensional Variational Inequalities and Complementarity Problems*. Springer-Verlag, 2003.

95. D. Challet, M. Marsili, and Y. C. Zhang, *Minority Games: Interacting Agents in Financial Markets*. Oxford University Press, 2005.
96. C. H. Yeung and Y. C. Zhang, "Minority games," in *Encyclopedia of Complexity and Systems Science*, R. A. Meyers, Ed., Springer-Verlag, pp. 1–26, 2013.
97. J. Parsons, *The Mobile Radio Propagation Channel*. John Wiley and Sons, 2000.
98. M. M. Buddhikot, "Understanding dynamic spectrum access: Models taxonomy, and challenges," in *Proceedings of the IEEE International Symposium on New Frontiers in Dynamic Spectrum Access Networks (DySPAN)*, 2007, pp. 1–14.
99. A. P. Subramanian and S. H. Gupta, "Fast spectrum allocation in coordinated spectrum access based cellular networks," in *Proceedings of the IEEE International Symposium on New Frontiers in Dynamic Spectrum Access Networks (DySPAN)*, 2007, pp. 320–330.
100. M. Wellens, J. Wu, and P. Mahonen, "Evaluation of spectrum occupancy in indoor and outdoor scenario in the context of cognitive radio," in *Proceedings of the International Conference on Cognitive Radio Oriented Wireless Networks and Communications (CROWNCOM)*, 2007, pp. 1–8.
101. D. J. Thomson, "Spectrum estimation and harmonic analysis," *Proceedings of the IEEE*, vol. 70, no. 9, pp. 1055–1096, 1982.
102. D. Slepian, "Prolate spheroidal wave functions, Fourier analysis and uncertainty," *Bell System Technical Journal*, vol. 57, pp. 1371–1430, 1978.
103. M. Frigo and S. G. Johnson, "The design and implementation of FFTW3," *Proceedings of the IEEE*, vol. 93, no. 2, pp. 216–231, 2005.
104. D. Percival and A. Walden, *Spectral Analysis for Physical Applications*. Cambridge University Press, 1993.
105. A. Drosopoulos and S. Haykin, "Angle-of-arrival estimation in the presence of multipath," in *Adaptive Radar Signal Processing*, S. Haykin, Ed., Springer-Verlag, pp. 11–90, 2007.
106. S. Haykin, *Adaptive Filter Theory*. Pearson, 5th ed., 2013.
107. S. Haykin, "Fundamental issues in cognitive radio," in *Cognitive Wireless Communication Networks*, E. Hossain and V. K. Bhargava, Eds., Springer-Verlag, pp. 1–43, 2007.
108. M. Loève, *Probability Theory*. Van Nostrand Reinhold Inc., 3rd revised ed., 1963.
109. S. Haykin and D. Thomson, "Signal detection in a nonstationary environment reformulated as an adaptive pattern classification problem," *Proceedings of the IEEE*, vol. 86, no. 11, pp. 2325–2344, 1998.
110. L. Cohen, *Time-Frequency Analysis*. Prentice-Hall, 1995.
111. L. Rayleigh, "On the spectrum of an irregular disturbance," *Philosophical Magazine*, vol. 41, pp. 238–243, 1903.
112. D. J. Thomson, "Multitaper analysis of nonstationary and nonlinear time series data," in *Nonlinear and Nonstationary Signal Processing*, W. Fitzgerald, R. Smith, A. Walden, and P. Young, Eds., Cambridge University Press, pp. 317–394, 2001.
113. B. Picinbono, "Second-order complex random vectors and normal distributions," *IEEE Transactions on Signal Processing*, vol. 44, no. 10, pp. 2637–2640, 1996.
114. C. N. K. Moores, "A technique for the cross-spectrum analysis of pairs of complex-valued time series, with emphasis on properties of polarized components and rotational invariants," *Deep Sea Research*, vol. 20, no. 12, pp. 1129–1141, 1973.
115. J. H. Middleton, "Outer rotary cross spectra, coherences, and phases," *Deep Sea Research Part A. Oceanographic Research Papers*, vol. 29, no. 10, pp. 1267–1269, 1982.

116. G. A. Prieto, F. L. Vernon, G. Masters, and D. J. Thomson, "Multitaper Wigner-Ville spectrum for detecting dispersive signals from earthquake records," in *Asilomar Conference on Signals, Systems, and Computers*, 2005, pp. 938–941.
117. W. A. Gardner, *Cyclostationarity in Communications and Signals Processing*. IEEE Press, 1994.
118. W. A. Gardner, *Statistical Spectral Analysis: A Nonprobabilistic Theory*. Prentice-Hall, 1988.
119. P. D. Sutton, K. E. Nolan, and L. E. Doyle, "Cyclostationary signatures in practical cognitive radio applications," *IEEE Journal on Selected Areas in Communications*, vol. 26, no. 1, pp. 13–24, 2008.
120. C. M. Spooner, *Theory and Application of Higher-Order Cyclostationarity*. Ph.D. Dissertation. University of California, Davis, CA, USA, 1992.
121. C. M. Spooner and W. A. Gardner, "The cumulant theory of cyclostationary time-series, Part II: development and applications," *IEEE Transactions on Signal Processing*, vol. 42, no. 12, pp. 3409–3429, 1994.
122. R. S. Roberts, W. A. Brown, and H. H. Loomism Jr., "Computationally efficient algorithms for cyclic spectral analysis," *IEEE Signal Processing Magazine*, vol. 8, no. 2, pp. 38–49, 1991.
123. W. C. Jakes, *Microwave Mobile Communications*. John Wiley and Sons, 1974.
124. T. Aulin, "A modified model for the fading signal at a mobile radio channel," *IEEE Transactions on Vehicular Technology*, vol. 28, no. 3, pp. 182–203, 1979.
125. J. F. Kaiser, "Nonrecursive digital filter design using the I_0(sinh) window function," in *Proceedings of the IEEE International Symposium on Circuits and Systems*, 1974, pp. 20–23.
126. S. O. Rice, "A short table of values of prolate spheroidal harmonics," *Bell Telephone Laboratories, MM 63-3241-13*, 1963.
127. B. Farhang-Boroujeny, "Filter bank spectrum sensing for cognitive radios," *IEEE Transactions on Signal Processing*, vol. 56, no. 5, pp. 1801–1811, 2008.
128. P. P. Vidyanathan, *Multirate Systems and Filter Banks*. Prentice-Hall, 1993.
129. W. Saad, Z. Han, M. Debbah, A. Hjörungnes, and T. Başar, "Coalitional games for distributed collaborative spectrum sensing in cognitive radio networks," in *Proceedings of IEEE INFOCOM*, 2009.
130. F. Khozeimeh and S. Haykin, "Brain-inspired dynamic spectrum management for cognitive radio Ad Hoc networks," *IEEE Transactions on Wireless Communications*, vol. 11, no. 10, pp. 3509–3517, 2012.
131. R. Diestel, *Graph Theory*. Springer-Verlag, 3rd ed., 2005.
132. D. Hebb, *The Organization of Behavior: A Neuropsychologocal Theory*. John Wiley and Sons, 1949.
133. D. Tsigankov and A. Koulakov, "A unifying model for activity-dependent and activity-independent mechanisms predicts complete structure of topographic maps in ephrin-a deficient mice," *Journal of Computational Neuroscience*, vol. 21, no. 1, pp. 101–114, 2006.
134. D. Willshaw and C. von der Malsburg, "A unifying model for activity-dependent and activity-independent mechanisms predicts complete structure of topographic maps in ephrin-a deficient mice," *Proceedings of the Royal Society of London Series B*, vol. 194, pp. 431–445, 1976.

135. Z. Quan, S. Cui, and A. H. Sayed, "Optimal linear cooperation for spectrum sensing in cognitive radio networks," *IEEE journal of Selected Topics in Signal Processing*, vol. 2, no. 1, pp. 28–40, 2008.

136. L. Wiskott and T. J. Sejnowski, "Constrained optimization for neural map formation: A unifying framework for weight growth and normalization," in *Self-Organizing Map Formation: Foundations of Neural Computation*, K. Obermayer and T. J. Sejnowski, Eds., The MIT Press, pp. 83–128, 2001.

137. D. Willkomm, S. Machiraju, J. Bolot, and A. Wolisz, "Primary user behavior in cellular networks and implications for dynamic spectrum access," *IEEE Communications Magazine*, vol. 47, no. 3, pp. 88–95, 2009.

138. T. Taleb, A. Jamalipour, Y. Nemoto, and N. Kato, "DEMAPS: A load-transition-based mobility management scheme for an efficient selection of map in mobile IPv6 networks," *IEEE Transactions on Vehicular Technology*, vol. 58, no. 2, pp. 954–965, 2008.

139. N. V. Lopes, M. J. Nicolau, and A. Santos, "Evaluating rate-estimation for a mobility and QoS-aware network architecture," in *Proceedings of the International Conference on Software*, 2009, *Telecommunications & Computer Networks (SoftCOM)*, pp. 348–352.

140. A.-C. Pang and H.-W. Tseng, "Dynamic backoff for wireless personal networks," in *Proceedings of the IEEE Global Telecommunications Conference (GLOBECOM)*, 2004, pp. 1580–1584.

141. Z. Wen, T. Luo, W. Xiang, S. Majhi, and Y. Ma, "Autoregressive spectrum hole prediction model for cognitive radio systems," in *Proceedings of the IEEE International Conference on Communications (ICC)*, 2008, pp. 254–257.

142. K. W. Choi and E. Hossain, "Opportunistic access to spectrum holes between packet bursts: a learning-based approach," vol. 10, no. 8, pp. 2497–2509, 2011.

143. D. Challet and Y. Zhang, "Emergence of cooperation and organization in an evolutionary game," *Physica A: Statistical and Theoretical Physic*, vol. 246, no. 3-4, pp. 407–418, 1997.

144. D. Challet, M. Marsili, and Y.-C. Zhan, *Minority Games: Interacting Agents in Financial Markets*. Oxford University Press, 2005.

145. P. Mahonen and M. Petrova, "Minority game for cognitive radios: Cooperating without cooperation," *Physical Communication*, vol. 1, no. 2, pp. 94–102, 2008.

146. M. Petrova, M. Michalopoulou, and P. M. Mahonen, "Self-organizing multiple access with minimal information: Networking in el farol bar," in *Proceedings of IEEE GLOBECOM Workshop*, 2010, pp. 1151–1156.

147. I. Aoki, K. Yamamoto, K. Kimura, H. Murata, and S. Yoshida, "Experimental study of minority game-based interference management for spectrum sharing," in *Proceedings of the IEEE International Conference on Communication Systems*, 2010, pp. 585–589.

148. Y. Saito, K. Yamamoto, H. Murata, and S. Yoshida, "Robust interference management to satisfy allowable outage probability using minority game," in *Proceedings of the IEEE International Symposium on Personal Indoor and Mobile Radio Communications*, 2010, pp. 2314–2319.

149. T. A. Weiss and F. K. Jondral, "Spectrum pooling: An innovative strategy for the enhancement of spectrum efficiency," *IEEE Communications Magazine*, vol. 42, no. 3, pp. S8–S14, 2004.

150. U. Berthold, F. K. Jondral, S. Brandes, and M. Schnell, "OFDM-based overlay systems: A promising approach for enhancing spectral efficiency," *IEEE Communications Magazine*, vol. 45, no. 12, pp. 52–58, 2007.

151. H. Arslan, H. A. Mahmoud, and T. Yucek, "OFDM for cognitive radio: Merits and challenges," in *Cognitive Radio, Software Defined Radio, and Adaptive Wireless Systems*, H. Arslan Ed., Springer-Verlag, 2007.

152. T. Keller and L. Hanzo, "Adaptive multicarrier modulation: A convenient framework for time-frequency processing in wireless communications," *Proceedings of the IEEE*, vol. 88, no. 5, pp. 611–640, 2000.

153. A. R. S. Bahai, B. R. Saltzberg, and M. Ergen, *Multi-Carrier Digital Communications: Theory and Applications of OFDM*. 2nd ed., Springer-Verlag, 2004.

154. K. E. Nolan, *Reconfigurable OFDM Systems*. Ph.D. Dissertation. University of Dublin, Trinity College, Ireland, 2005.

155. Y. Li and G. Stuber, *Orthogonal Frequency Division Multiplexing for Wireless Communications*. Springer-Verlag, 2006.

156. M. Sterand, T. Svensson, T. Ottosson, A. Ahlen, A. Svensson, and A. Brunstrom, "Towards systems beyond 3G based on adaptive OFDMA transmission," *Proceedings of the IEEE*, vol. 95, no. 12, pp. 2432–2455, 2007.

157. M. Morelli, C. C. J. Kuo, and M. O. Pun, "Synchronization techniques for orthogonal frequency division multiple access (OFDMA): A tutorial review," *Proceedings of the IEEE*, vol. 95, no. 7, pp. 1394–1427, 2007.

158. A. Ghosh, J. Zhang, J. G. Andrews, and R. Muhamed, *Fundamentals of LTE*. Pearson Education, 2010.

159. B. Farhang-Boroujeny, "OFDM versus filter bank multicarrier," *IEEE Signal Processing Magazine*, vol. 28, no. 3, pp. 92–112, 2011.

160. N. Benvenuto, R. Dinis, D. Falconer, and S. Tomasin, "Single carrier modulation with nonlinear frequency domain equilization: An idea whose time has come again," *Proceedings of the IEEE*, vol. 98, no. 1, pp. 69–96, 2010.

161. P. Banelli, S. Buzzi, G. Colavolpe, A. Modenini, F. Rusek, and A. Ugolini, "Modulation formats and waveforms for 5G networks: Who will be the heir of OFDM? An overview of alternative modulation schemes for improved spectral efficiency," *IEEE Signal Processing Magazine*, vol. 31, no. 6, pp. 80–93, 2014.

162. D. Wübben, P. Rost, J. Bartlet, M. Lalam, V. Savin, M. Gorgoglione, A. Dekorsy, and G. Fettweis, "Benefits and impact of cloud computing on 5G signal processing," *IEEE Signal Processing Magazine*, vol. 31, no. 6, pp. 35–44, 2014.

163. A. van der Schaft and H. Schumacher, *An Introduction to Hybrid Dynamical Systems*. Springer-Verlag, 2000.

164. R. Goebel, R. G. Sanfelice, and A. R. Teel, "Hybrid dynamical systems," *IEEE Control Systems Magazine*, vol. 29, no. 2, pp. 28–93, 2009.

165. D. P. Bertsekas and J. N. Tsitsiklis, *Parallel and Distributed Computation: Numerical Methods*. Prentice-Hall, 1989.

166. R. D. Yates, "A framework for uplink power control in cellular radio systems," *IEEE Journal on Selected Areas in Communications*, vol. 13, no. 7, pp. 1341–1347, 1995.

167. G. Scutari, D. P. Palomar, and S. Barbarossa, "Simultaneous iterative water-filling for Gaussian frequency-selective interference channels," in *Proceedings of the IEEE International Symposium on Information Theory*, 2006, pp. 600–604.

168. G. Scutari, D. P. Palomar, and S. Barbarossa, "Asynchronous iterative waterfilling for Gaussian frequency-selective interference channels: A unified framework," in *Proceedings of the Information Theory and Applications Workshop*, 2007, pp. 349–358.
169. I. Konnov, *Equilibrium Models and Variational Inequalities*. Springer-Verlag, 2007.
170. M. Biagi, *Cross-Layer Optimization of Multi-Antenna 4G-WLANs*. Ph.D. Dissertation. University of Rome La Sapienza, Rome, Italy, 2005.
171. N. Nei and C. Comaniciu, "Adaptive channel allocation spectrum ettiquette for cognitive radio networks," in *Proceedings of the 1st IEEE International Symposium on New Frontiers in Dynamic Spectrum Access Networks (DySPAN)*, 2005, pp. 269–278.
172. J. O. Neel, *Analysis and Design of Cognitive Radio Networks and Distributed Radio Resource Management Algorithms*. Ph.D. Dissertation. Virginia Polytechnic Institute and State University, 2006.
173. C. Liang, *A Game-Theoretic Approach to Power Management in MIMO-OFDM Ad Hoc Networks*. Ph.D. Dissertation. Drexel University, 2006.
174. M. Felegyhazi, *Non-Cooperative Behavior in Wireless Networks*. Ph.D. Dissertation. Ecole Polytechnique Federale De Lausanne (EPFL), Switzerland, 2007.
175. M. Felegyhazi, M. Cagalj, S. S. Bidokhti, and J. P. Hubaux, "Non-cooperative multi-radio channel allocation in wireless networks," in *Proceedings of the IEEE International Conference on Computer Communications (INFOCOM)*, 2007, pp. 1442–1450.
176. F. Wang, M. Krunz, and S. Cui, "Price-based spectrum management in cognitive radio networks," *IEEE Journal of Selected Topics in Signal Processing*, vol. 2, no. 1, pp. 74–87, 2008.
177. G. Scutari, D. P. Palomar, J. S. Pang, and F. Facchinei, "Flexible design of cognitive radio wireless systems," *IEEE Signal Processing Magazine*, vol. 26, no. 5, pp. 107–123, 2009.
178. C. E. Shannon and W. Weaver, *The Mathematical Theory of Communication*. University of Illinois Press, 1998.
179. T. M. Cover and J. A. Thomas, *Elements of Information Theory*. 2nd ed., Wiley-Interscience, 2006.
180. S. Haykin, *Communication Systems*. John Wiley and Sons, 4th ed., 2001.
181. W. Yu, *Competetion and Cooperation in Multi-user Communication Environments*. Ph.D. Dissertation. Stanford University, USA, 2002.
182. A. B. Carleial, "Interference channels," *IEEE Transactions on Information Theory*, vol. 24, no. 1, pp. 60–70, 1978.
183. Z. Q. Luo and S. Zhang, "Dynamic spectrum management: Complexity and duality," *IEEE Journal of Selected Topics in Signal Processing*, vol. 2, no. 1, pp. 57–73, 2008.
184. S. Hayashi and Z. Q. Luo, "Spectrum management for interference-limited multiuser communication systems," *IEEE Transactions on Information Theory*, vol. 55, no. 3, pp. 1153–1175, 2009.
185. N. Yamashita and Z. Q. Luo, "A nonlinear complementarity approach to multiuser power control for digital subscriber lines," *Optimization Methods and Software*, vol. 19, no. 5, pp. 633–652, 2004.
186. Z. Q. Luo, J. S. Pang, and D. Ralph, *Mathematical Programs with Equilibrium Constraints*. Cambridge University Press, 1996.
187. Z. Q. Luo and J. S. Pang, "Analysis of iterative waterfilling algorithm for multiuser power control in digital subscriber lines," *EURASIP Journal on Applied Signal Processing*, vol. 2006, pp. 1–10, 2006.

188. W. Yu, G. Ginis, and J. M. Cioffi, "Distributed multiuser power control for digital subscriber lines," *IEEE Journal on Selected Areas in Communications*, vol. 20, no. 5, pp. 1105–1115, 2002.

189. J. B. Rosen, "Existence and uniqueness of equilibrium points for concave n-person games," *Econometrica*, vol. 33, pp. 520–534, 1965.

190. K. J. Arrow and G. Debreu, "Existence of an equilibrium for a competitive economy," *Econometrica*, vol. 22, pp. 265–290, 1954.

191. F. Facchinei and C. Kanzow, "Generalized Nash equilibrium problems," *4OR: A Quarterly Journal of Operations Research*, vol. 5, no. 3, pp. 173–210, 2007.

192. J.-P. Aubin, *Mathematical Methods of Game and Economic Theory*. North-Holland Publishing Company, Revised Edition, 1982.

193. G. Scutari, D. P. Palomar, and S. Barbarossa, "Optimal linear precoding strategies for wideband noncooperative systems based on game theory-Part I: Nash equilibria," *IEEE Transactions on Signal Processing*, vol. 56, no. 3, pp. 1230–1249, 2008.

194. G. Hardin, "The tragedy of the commons," *Science*, vol. 162, no. 3859, pp. 1243–1248, 1968.

195. M. Haddad, S. E. Elayoubi, E. Altman, and Z. Altman, "A hybrid approach for radio resource management in heterogeneous cognitive networks," *IEEE Journal on Selected Areas in Communications*, vol. 29, no. 4, pp. 831–842, 2011.

196. Y. Liu and L. Dong, "Spectrum sharing in MIMO cognitive radio networks based on cooperative game theory," *IEEE Transactions on Wireless Communications*, vol. 13, no. 9, pp. 4807–4820, 2014.

197. M. Yang, T. Groves, N. Zheng, and P. Cosman, "Iterative pricing-based rate allocation for video streams with fluctuating bandwidth availability," *IEEE Transactions on Multimedia*, vol. 16, no. 7, pp. 1849–1862, 2014.

198. K. B. Song, S. T. Chung, G. Ginis, and J. M. Cioffi, "Dynamic spectrum management for next-generation DSL systems," *IEEE Communications Magazine*, vol. 40, no. 10, pp. 101–109, 2002.

199. S. T. Chung, S. J. Kim, J. Lee, and J. M. Cioffi, "A game theoretic approach to power allocation in frequency-selective Gaussian interference channels," in *Proceedings of the IEEE International Symposium on Information Theory (ISIT)*, 2003, pp. 316–316.

200. F. Khozeimeh and S. Haykin, "Self-organizing dynamic spectrum management for cognitive radio networks," in *Proceedings of the 8th Conference on Communication Networks and Services Research (CNSR)*, 2010, pp. 1–7.

201. M. Aghassi and D. Bertsimas, "Robust game theory," *Mathematical Programming Series B*, vol. 107, pp. 231–273, 2006.

202. M. K. Brady, *Techniques for Interference Analysis and Spectrum Management of Digital Subscriber Lines*. Ph.D. Dissertation. University of California, Berkeley, CA, USA, 2006.

203. K. M. Teo, *Nonconvex Robust Optimization*. Ph.D. Dissertation, Massachusets Institute of Technology, USA, 2007.

204. R. H. Gohary, Y. Huang, Z. Q. Luo, and J. S. Pang, "A generalized iterative water-filling algorithm for distributed power control in the presence of a jammer," *IEEE Transactions on Signal Processing*, vol. 57, no. 7, pp. 2660–2674, 2009.

205. D. J. Love, R. W. Heath, W. Santipach, and M. L. Honig, "What is the value of limited feedback for MIMO channels?" *IEEE Communications Magazine*, vol. 42, no. 10, pp. 54–59, 2004.

206. D. J. Love, R. W. Heath, V. K. N. Lau, D. Gesbert, B. D. Rao, and M. Andrews, "An overview of limited feedback in wireless communication systems," *IEEE Journal on Selected Areas in Communications*, vol. 26, no. 8, pp. 1341–1365, 2008.

207. C. E. Shannon, "A mathematical theory of communication," *The Bell System Technical Journal*, vol. 27, pp. 379–423, 623–656, 1948.

208. R. A. Howard, "Information value theory," *IEEE Transactions on Systems Science and Cybernetics*, vol. SSC-2, no. 1, pp. 22–26, 1966.

209. P. Corning, "Control information: The missing element in Norbert Wiener's cybernetic paradigm," *Kybernetics*, vol. 30, no. 9-10, pp. 1272–1288, 2001.

210. N. Tishby and D. Polani, "Information theory of decisions and actions," in *Perception-Action Cycle: Models, Architectures, and Hardware*, V. Cutsuridis, A. Hussain, and J. G. Taylor, Springer-Verlag, pp. 601–636, 2011.

211. P. Setoodeh, S. Haykin, and K. Rezaei-Moghadam, "Double-layer dynamics of cognitive radio networks," in *Proceedings of the IEEE International Workshop on Emerging Cognitive Radio Applications and Algorithms (CORAL)*, 2012, pp. 1–6.

212. K. Zhu, D. Niyato, P. Wang, and Z. Han, "Dynamic spectrum leasing and service selection in spectrum secondary market of cognitive radio networks," *IEEE Transactions on Wireless Communications*, vol. 11, no. 9, pp. 1136–1145, 2012.

213. L. Marsh and C. Onof, "Stigmergic epistemology, stigmergic cognition," *Cognitive Systems Research*, vol. 9, no. 1-2, pp. 136–148, 2008.

214. A. Stephan, "The dual role of emergence in the philosophy of mind and in cognitive science," *Synthese*, vol. 151, pp. 485–498, 2006.

215. D. C. Tarraf and H. H. Asada, "Decentralized hierarchical control of multiple time scale systems," in *Proceedings of the American Control Conference*, 2002, pp. 1121–1122.

216. N. D. Duong, A. S. Madhukumar, and D. Niyato, "Stackelberg Bayesian game for power allocation in two-tier networks," *IEEE Transactions on Vehicular Technology*, vol. 65, no. 4, pp. 2341–2354, 2016.

217. D. Mignone, *Control and Estimation of Hybrid Systems with Mathematical Optimization*. Ph.D. Dissertation. Swiss Federal Institute of Technology (ETH), Switzerland, 2002.

218. T. Steffen, *Control Reconfiguration of Dynamical Systems*. Springer-Verlag, 2005.

219. M. L. Aghassi, *Robust Optimization, Game Theory, and Variational Inequalities*. Ph.D. Dissertation. Massachusetts Institute of Technology (MIT), 2005.

220. M. Fazel, D. M. Gayme, and M. Chiang, "Transient analysis for wireless power control," in *Proceedings of the IEEE Global Telecommunications Conference (GLOBECOM)*, 2006, pp. 1–6.

221. A. Nagurney and D. Zhang, *Projected Dynamical Systems and Variational Inequalities with Applications*. Springer-Verlag, 1996.

222. S. Skogestad and I. Postlethwaite, *Multivariable Feedback Control: Analysis and Design*. John Wiley and Sons, 2nd ed., 2006.

223. M. G. Cojocaru, P. Daniele, and A. Nagurnay, "Double-layered dynamics: A unified theory of projected dynamical systems and evolutionary variational inequalities," *European Journal of Operational Research*, vol. 175, pp. 494–507, 2006.

224. J. Yang, A. G. Klein, and D. R. BrownIII, "Natural cooperation in wireless networks," *IEEE Signal Processing Magazine*, vol. 26, no. 5, pp. 98–106, 2009.

225. F. J. Christophersen, *Optimal Control of Constrained Piecewise Affine Systems*. Springer-Verlag, 2007.

226. J. Roll, *Local and Piecewise Affine Approaches to System Identification*. Ph.D. Dissertation. Linköping University, Linköping, Sweden, 2003.

227. L. Engelson, "On dynamics of traffic queues in a road network with route choice based on real time traffic information," *Transportation Research Part C*, vol. 11, pp. 161–183, 2003.

228. V. B. Kolmanovskii and V. R. Nosov, *Stability of Functional Differential Equations*. Academic Press, 1986.

229. J. K. Hale and S. M. V. Lunel, *Introduction to Functional Differential Equations*. Springer-Verlag, 1993.

230. V. B. Kolmanovskii and A. D. Myshkis, *Introduction to the Theory and Applications of Functional Differential Equations*. Kluwer Academic Publishers, 1999.

231. L. E. El'sgol'ts and S. B. Norkin, *Introduction to the Theory and Application of Differential Equations with Deviating Arguments*. Academic Press, 1973.

232. S. I. Niculescu, C. E. de Souza, L. Dugard, and J. M. Dion, "Robust exponential stability of uncertain systems with time-varying delays," *IEEE Transactions on Automatic Control*, vol. 43, no. 5, pp. 743–748, 1998.

233. B. O. Palsson, *Systems Biology: Simulation of Dynamic Network States*. Cambridge University Press, 2011.

234. N. Hadjisavvas and S. Schaible, "Generalized monotone maps," in *Handbook of Generalized Convexity and Generalized Monotonicity*, N. Hadjisavvas, S. Kmlosi, and S. Schaible, Eds., Springer-Verlag, pp. 389–420, 2005.

235. M. C. Vuran and I. F. Akyildiz, "A-MAC: Adaptive medium access control for next generation wireless terminals," *IEEE/ACM Transactions on Networking*, vol. 15, no. 3, pp. 574–587, 2007.

236. K. Zhu, D. Niyato, P. Wang, and Z. Han, "Dynamic spectrum leasing and service selection in spectrum secondary market of cognitive radio networks," *IEEE Transactions on Wireless Communications*, vol. 11, no. 3, pp. 1136–1145, 2012.

237. V. Grover and M. K. Malhotra, "Transaction cost framework in operations and supply chain management research: Theory and measurement," *Journal of Operations Management*, vol. 21, no. 4, pp. 457–473, 2003.

238. E. Estrada, *The Structure of Complex Networks*. Oxford University Press, 2012.

239. W. Saad, Z. Han, T. Başar, M. Debbah, and A. Hjørungnes, "Coalition formation games for collaborative spectrum sensing," *IEEE Transactions on Vehicular Technology*, vol. 60, no. 1, pp. 276–297, 2011.

240. D. Li, Y. Xu, X. Wang, and M. Guizani, "Coalitional game theoretic approach for secondary spectrum access in cooperative cognitive radio networks," *IEEE Transactions on Wireless Communications*, vol. 10, no. 3, pp. 844–856, 2011.

241. F. Ciucci, T. Honda, and M. C. Yang, "An information-passing strategy for achieving Pareto optimality in the design of complex systems," *Research in Engineering Design*, vol. 23, no. 1, pp. 71–83, 2012.

242. D. D. Wu, C. Luo, and D. L. Olson, "Efficiency evaluation for supply chains using maximin decision support," *IEEE Transactions on Systems, Man, and Cybernetics: Systems*, vol. 44, no. 8, pp. 1088–1097, 2014.

243. C. Y. Lee and A. L. Johnson, "Measuring efficiency in imperfectly competitive markets: An example of rational inefficiency," *Journal of Optimization Theory and Applications*, vol. 164, no. 2, pp. 702–722, 2015.

244. M. F. Balcan, F. Constantin, G. Piliouras, and J. S. Shamma, "Game couplings: Learning dynamics and applications," in *Proceedings of the IEEE Conference on Decision and Control and European Control Conference (CDC-ECC)*, 2011, pp. 2441–2446.
245. T. Honda, F. Ciucci, K. E. Lewis, and M. C. Yang, "Comparison of information passing strategies in system-level modeling," *AIAA Journal*, vol. 53, no. 5, pp. 1121–1133, 2015.
246. H. B. Yilmaz, T. Tugcu, F. Alagöz, and S. Bayhan, "Radio environment map as enabler for practical cognitive radio networks," *IEEE Communications Magazine*, vol. 51, no. 12, pp. 162–169, 2013.
247. S. Barbarossa, S. Sardelliti, and P. Di Lorenzo, "Communicating while computing: Distributed mobile cloud computing over 5G heterogeneous networks," *IEEE Signal Processing Magazine*, vol. 31, no. 6, pp. 45–55, 2014.
248. M. El-Refaey, N. Magdi, and H. Abd El-Megeed, "Cloud-assisted spectrum management system with trading engine," in *Proceedings of the IEEE International Conference on Wireless Communications and Mobile Computing (IWCMC)*, 2014, pp. 953–958.
249. P. C. Jain, "Rural wireless broadband internet access in wireless regional area network using cognitive radio," in *Proceedings of the IEEE International Conference on Signal Processing and Communication (ICSC)*, 2013, pp. 98–103.
250. J. Hecht, "The bandwidth bottleneck," *Nature*, vol. 536, pp. 139–142, 2016.
251. A. B. Atkinson and J. E. Stiglitz, *Lectures on Public Economics*. Princeton University Press, 2015.
252. M. Van Essen and M. Walker, "A simple market-like allocation mechanism for public goods," *Games and Economic Behavior*, vol. 101, pp. 6–19, 2017.
253. E. Lindahl, "Just taxation-a positive solution," in *Classics in the Theory of Public Finance*, R. Musgrave and A. Peacock, Eds., Academic Press, pp. 98–123, 1958.
254. P. Samuelson, *Foundations of Economic Analysis*. Harvard University Press, 1983.
255. T. Bergstrom, "Collective choice and the Lindahl allocation method," in *Theory and Measurement of Economic Externalities*, S. A. Y. Lin, Eds., Academic Press, pp. 107–131, 1976.
256. S. Haykin and P. Setoodeh, "Cognitive heterogeneous networks: Economic provisioning for 5G," *IEEE COMSOC TCCN Communications*, vol. 1, no. 1, pp. 6–9, 2015.
257. A. Nagurney, "Optimal supply chain network design and redesign at minimal total cost and with demand satisfaction," *International Journal of Production Economics*, vol. 28, no. 1, pp. 200–208, 2010.
258. T. Bergstrom, *Theory of Public Goods and Externalities*. Lecture Notes, University of California, Santa Barbara, CA.
259. S. Landström, A. Furuskar, K. Johansson, L. Falconetti, and F. Kronestedt, "Heterogeneous networks-increasing cellular capacity," *Ericsson Review*, pp. 1–6, 2011.
260. Y. L. Lee, T. C. Chuah, J. Loo, and A. Vinel, "Recent advances in radio resource management for heterogeneous LTE/LTE-A networks," *IEEE Communications Surveys & Tutorials*, vol. 16, no. 4, pp. 2142–2180, 2014.
261. J. G. Andrews, S. Buzzi, C. Wan, S. V. Hanly, A. Lozano, A. C. K. Soong, and J. C. Zhang, "What will 5G be?" *IEEE Journal on Selected Areas in Communications*, vol. 32, no. 6, pp. 1065–1082, 2014.
262. E. Hossain and M. Hasan, "5G cellular: Key enabling technologies and research challenges," *IEEE Instrumentation & Measurement Magazine*, vol. 18, no. 3, pp. 11–21, 2015.

263. T. S. Rappaport, S. Sun, R. Mayzus, H. Zhao, Y. Azar, K. Wang, G. N. Wong, J. K. Schulz, M. Samimi, and F. Gutierrez, "Millimeter wave mobile communications for 5G cellular: It will work!" *IEEE Access*, vol. 1, pp. 335–349, 2013.

264. E. G. Larsson, O. Edfors, F. Tufvesson, and T. L. Marzetta, "Massive MIMO for next generation wireless systems," *IEEE Communications Magazine*, vol. 52, no. 2, pp. 186–195, 2014.

265. A. Damnjanovic, J. Montojo, Y. Wei, T. Ji, T. Luo, M. Vajapeyam, T. Yoo, O. Song, and D. Malladi, "A survey on 3GPP heterogeneous networks," *IEEE Wireless Communications*, vol. 18, no. 3, pp. 10–21, 2011.

266. I. Hwang, B. Song, and S. S. Soliman, "A holistic view on hyper-dense heterogeneous and small cell networks," *IEEE Communications Magazine*, vol. 51, no. 6, pp. 20–27, 2013.

267. J. G. Andrews, "Seven ways that HetNets are a cellular paradigm shift," *IEEE Communications Magazine*, vol. 51, no. 3, pp. 136–144, 2013.

268. C. Fang, F. R. Yu, T. Huang, J. Liu, and Y. Liu, "A survey of green information-centric networking: Research issues and challenges," *IEEE Communications Surveys & Tutorials*, pp. 1–19, 2015.

269. S. J. Daraka, S. Dhabub, C. Moyc, H. Zhanga, J. Palicotc, and A. P. Vinod, "Low complexity and efficient dynamic spectrum learning and tunable bandwidth access for heterogeneous decentralized cognitive radio networks," *Digital Signal Processing*, vol. 37, pp. 13–23, 2015.

270. M. Mustonen, M. Matinmikko, M. Palola, S. Yrjölä, and K. Horneman, "An evolution toward cognitive cellular systems: Licensed shared access for network optimization," *IEEE Communications Magazine*, vol. 53, no. 5, pp. 68–74, 2015.

271. F. H. Panahi and T. Ohtsuki, "Stochastic geometry modeling and analysis of cognitive heterogeneous cellular networks," *EURASIP Journal on Wireless Communications and Networking*, vol. 2015, no. 1, Article No. 141, 2015.

272. R. Berry, M. L. Honig, and R. Vohra, "Spectrum markets: Motivation, challenges, and implications," *IEEE Communications Magazine*, vol. 48, no. 11, pp. 146–155, 2010.

273. R. Jain, "Network market design Part I: Bandwidth markets," *IEEE Communications Magazine*, vol. 50, no. 11, pp. 78–83, 2012.

274. R. A. Berry, "Network market design Part II: Spectrum markets," *IEEE Communications Magazine*, vol. 50, no. 11, pp. 84–90, 2012.

275. M. Hannebauer, *Autonomous Dynamic Reconfiguration in Multi-Agent Systems: Improving the Quality and Efficiency of Collaborative Problem Solving*. Springer, 2002.

276. A. Nagurney, "A system-optimization perspective for supply chain network integration: The horizontal merger case," *Transportation Research Part E: Logistics and Transportation Review*, vol. 45, no. 1, pp. 1–15, 2009.

277. S. Boyd and L. Vandenberghe, *Convex Optimization*. Cambridge University Press, 2004.

278. M. S. Bazaraa, H. D. Sherali, and C. M. Shetty, *Nonlinear Programming: Theory and Algorithms*. Wiley-Interscience, 3rd ed., 2006.

279. D. P. Bertsekas, A. Nedic, and A. Ozdaglar, *Convex Analysis and Optimization*. Athena Scientific, 2003.

280. R. W. Cottle, J. S. Pang, and R. E. Stone, *The Linear Complementarity Problem*. Academic Press, 1992.

281. M. G. Cojocaru, *Projected Dynamical Systems on Hilbert Spaces*. Ph.D. Dissertation. Queen's University, Canada, 2002.

282. M. G. Cojocaru and L. B. Jonker, "Existence of solutions to projected differential equations in Hilbert spaces," *Proceedings of the American Mathematical Society*, vol. 132, no. 1, pp. 183–193, 2004.

283. A. V. Skorokhod, "Stochastic equations for diffusions in a bounded region," *Theory of Probability and its Applications*, vol. 6, pp. 264–274, 1961.

284. V. I. Rozhkov and A. M. Popov, "Inequalities for solutions of certain systems of differential equations with large time-lag," *Differential Equations*, vol. 7, no. 2, pp. 271–278, 1971.

285. M. G. Cojocaru, P. Daniele, and A. Nagurney, "Projected dynamical systems, evolutionary variational inequalities, applications, and a computational procedure," in *Pareto Optimality, Game Theory, and Equilibria*, A. Chinchuluun, P. M. Paradalos, A. Migdalas, and L. Pitsoulis, Eds., Springer-Verlag, pp. 387–406, 2008.

INDEX

Fundamentals of Cognitive Radio, First Edition. Peyman Setoodeh and Simon Haykin.